Agents of
INNOVATION

Agents of
INNOVATION
The General Board and the Design of the Fleet
That Defeated the Japanese Navy

by John T. Kuehn

NAVAL INSTITUTE PRESS
Annapolis, Maryland

This book has been brought to publication with the generous assistance of Edward S. and Joyce I. Miller.

Naval Institute Press
291 Wood Road
Annapolis, MD 21402

© 2008 by John T. Kuehn
All rights reserved. No part of this book may be reproduced or utilized in any form or by any means, electronic or mechanical, including photocopying and recording, or by any information storage and retrieval system, without permission in writing from the publisher.

Library of Congress Cataloging-in-Publication Data

Kuehn, John T.
  Agents of innovation : the General Board and the design of the fleet that defeated the Japanese / John T. Kuehn.
    p. cm.
  Includes bibliographical references and index.
  ISBN 978-1-59114-448-9 (alk. paper)
  1. Sea-power—United States—History—20th century. 2. United States Navy—Reserve fleets—History—20th century. 3. United States. Navy General Board—History. 4. Warships—Technological innovations—United States—History—20th century. 5. World War, 1939–1945—Naval operations, American. I. Title.
  VA58.K84 2008
  359'.03097309041—dc22
                                                                                    2008015597

Printed in the United States of America on acid-free paper ♾

14 13 12 11 10 09 08   9 8 7 6 5 4 3 2
First printing

Contents

List of Illustrations		vii
Acknowledgments		ix
Preface		xiii
List of Acronyms		xviii
Chapter 1.	Introduction: The Navy, Treaties, and Innovation	1
Chapter 2.	The General Board	8
Chapter 3.	U.S. Sea Power and the Washington Conference	23
Chapter 4.	The General Board and the Treaty System	40
Chapter 5.	Battleship Modernization	63
Chapter 6.	Naval Aviation and the Fortification Clause	88
Chapter 7.	Strategic Innovation in the Interwar U.S. Navy	125
Chapter 8.	Perspectives from Great Britain, Japan, and Germany	144
Chapter 9.	Conclusion	162
Appendix 1.	The Washington Naval Treaty	188
Appendix 2.	U.S. Naval Policy, 1922	198
Appendix 3.	Bureau Recommendations for Battleship Modernization	206
Appendix 4.	Comparison Chart for Cruiser Designs	208
Appendix 5.	Excerpt from the Mobile Base Project	210
Notes		213
Bibliography		249
Index		257

Illustrations

Figures

Figure 1. Navy Organizational Relationships during the Interwar Period   13
Figure 2. Flying Deck Cruiser Design Blueprint   120

Map

Map 1. Map of the Pacific   129

Tables

Table 1. Innovation Relationships   5
Table 2. Initial Design Proposal for Flying Deck Cruiser by the Bureau of Construction and Repair   111
Table 3. Initial Design Proposal for Flying Deck Cruiser by the Bureau of Aeronautics   112
Table 4. Characteristics for Flying Deck Cruiser by CNO and the General Board   121

Acknowledgments

There are so many folks involved in the genesis and completion of a project like this that to write an acknowledgment or dedication is a somewhat daunting task. Nonetheless, and at the risk of being verbose (guaranteed), here goes. First, all errors, whether factual, interpretative, or accidental are mine and mine alone.

The book you hold began as a doctoral dissertation. That said, I believe the most important and influential folks who motivated and encouraged me should be up front—Kimberlee Wade Kuehn and Donald J. Mrozek. If there truly are no such things as coincidences, then God truly blessed me in sending these two into my life. Kimberlee, who I have know for twenty-seven years, and been married to for twenty-five, could have been a show stopper but instead became chief cheerleader, never flagging in her support. After I obtained my third masters degree she could have simply said, "Enough!" Instead, when Kansas State University and the Army Command and General Staff College first began a trial association for a military history Ph.D. relationship, she encouraged me to get in on the ground floor.

Enter Don Mrozek, who was my first professor for a seminar in topics related to American military history. Don took me under his wing and encouraged and guided me at every step. If not for Kimberlee and Don I would not be writing and teaching history today. To Don must also go the blame for my topic. I wanted to write about Napoleonic coalition warfare, but Don suggested my knowledge of naval operations, especially in the Pacific, as well as my real lack of skill with European languages, probably argued for contemporary research in American/ English-language sources on an American topic. It was also Don who steered me away from Europe and back to Asia and the Pacific. Having spent almost nine years of my life overseas in Japan, Guam, and the Philippines, this, in retrospect, made superb sense.

Why the interwar U.S. Navy and the Washington Naval Treaty? Chris Gabel, longtime professor at the U.S. Army Command and General Staff is guilty in regard to the specific topic. Chris has long taught innovation during the period between World Wars I and II and I, having taken his course in its long version, became fascinated with interwar dynamics of navies and arms treaties. The other members of my dissertation committee also provided invaluable aid and assistance: David Stone, David Graff, and Dale Herspring. Each of them brought a sensibility to the complex topic that was and is greatly appreciated.

The research I did on the topic was stimulated by another series of fortunate encounters. The folks at the Hoover Library, especially Spencer Howard, were very helpful in introducing me to the wonderful material there on arms control in the 1920s and 1930s and the role of the General Board in naval policymaking and strategy. At the 2004 Naval Institute/Naval Historical Center history symposium I found myself sitting next to a National Archives and Records Administration (NARA) gent named Mark Mollen. I shared my project plans with him and he told me to get ahold of him if I visited Washington, D.C. during my research trips. When I had cause to be stationed in the Washington area for several months on a teaching assignment, I sent Mark an email and he graciously met me and showed me the ropes at the downtown Archives in D.C. There he introduced me to two other archivists were also invaluable in helping me navigate the studies of the General Board of the Navy: Rebecca Livingstone and Charles Johnson. At the College Park archives I am indebted to Barry Zerby for making available the advanced base studies and 1924 War Plan Orange from the OpNav archives. Barry is a national treasure. The Naval Historical Center is another invaluable resource in the D.C. area and to Jeffrey Barlow goes all the credit for treating me like visiting royalty as I used their excellent archives and the superb Navy Department Library. It was here that Cathy Lloyd introduced me to the Proceedings and Hearings of the General Board which really opened up the world of the interwar Navy to me, I cannot thank her and Jeff enough. Dan Kuehl at National Defense University was also a key supporter during my time in Washington.

There have been a number of other folks who have supported my efforts, both online and in person. H. P. "Ned" Willmott, Timothy Francis, Jon Parshall, John Lundstrom, Mark Mandeles, and, especially, Sadao Asada provided voluminous advice via email correspondence and it is my dearest wish that I may someday meet these gentlemen in the flesh to personally thank them for their support and advice. Three gents who I did finally meet after email interaction were David Ulbrich, Will O'Neil, and Thomas C. Hone. David Ulbrich was especially helpful as I reviewed the literature on the topic and Will O'Neil, who specializes in the organizational dynamics of the interwar period, has been a valuable critic throughout. My association to the Hone family came about as a result of a chance meeting with Trent Hone at Annapolis and Trent provided excellent feedback and advice and it was through him that I had cause to interact with his father, Tom. Finally,

Edward Miller, author of the superlative *War Plan Orange*, was instrumental in my seeing the project to completion during the endgame and in turning a turgid and lengthy dissertation into a readable naval history.

Finally, Dennis Giangreco, my long-time associate and sometime co-author (on other projects), has been a wonderful source of advice, inspiration, and encouragement throughout the entire project. I want to say the same for my former Navy submarine officer brother, Robert B. Kuehn, who is probably one of the few people other than my committee to have read through the entire dissertation. He suffered innumerable emails during the project. I am sure there are others who have assisted beyond those named here and to them I send out a heartfelt thank you as well.

Preface

This book has a very simple premise—the U.S. Navy's contribution to victory in the Pacific during World War II can be understood only by studying how the General Board of the U.S. Navy constructed the "treaty navy" during the period between the wars. The naval arms limitation system as implemented by the General Board was a positive contributing factor to the success of the Navy when it went to war in the Pacific in December 1941—perhaps more than if no naval limitation had been implemented at all. The traditional view of the period, advanced by the esteemed Samuel Eliot Morison, is that World War II found the U.S. Navy less ready than it might have been had the United States not subscribed to a regime of naval limitation treaties. Morison and his heirs have argued that these treaties materially and tactically put the United States at a severe disadvantage at the outset of war. The United States, they argue, would have been better served to have refrained from participation in naval arms limitation. Also, nowhere do these accounts articulate the role of the General Board. This view has muddied the water for over half a century.

Much of the focus of the arms limitation treaties has rightly focused on how they directly affected naval construction. A capital ship "holiday" froze battleship technology in place, placed limits on aircraft carrier construction, and imposed tonnage limitations that prevented navies from building any surface combatants larger than 10,000-ton cruisers for nearly fifteen years. The overall tonnage of capital ships was limited by a set ratio for the five principal naval powers—the United States, Great Britain, Japan, Italy, and France. However, contemporary U.S. Navy observers of the period were far more upset about the compromise that the U.S. delegation at the Washington Naval Conference accepted that led to the initiation of the whole system in the first place—Article XIX. This article established a status quo for naval fortification in the western Pacific that drastically altered the way that the U.S. Navy viewed the application of sea power. U.S. Navy Capt. Dudley W. Knox scathingly identified the fortification clause as "Of even greater importance than the loss to us in tonnage strength. . . ."[1] Ironically, this clause forced the U.S. Navy to choose between abandoning its existing strategy for protecting U.S. interests in the Far East or to come up with a modified conception of sea power. To its great credit the U.S. Navy, principally through the efforts of the General Board, chose the latter course. Implicit in the traditional view of the U.S. Navy of this period is that it was composed predominately of hide-bound, reactionary battleship admirals whose minds were closed to innovation. It is often forgotten that the "dreadnought" admirals were the innovators of their day, that they had often been the "wild-eyed" impassioned agents of change prior to assuming leadership roles in the interwar Navy. Accordingly, I decided to focus on what has often been mischaracterized as the icon of conservatism and inflexibility in the U.S. Navy of the period—the General Board.

The history of the United States prior to World War II contains few untold stories of great importance to the Navy as an institution. The General Board as an organization is little-known, and for good reason. A small organization of never more than about twelve officers, mostly captains or admirals, the General Board of the Navy was conceived and created in 1900 with the view that it would serve as a sort of Naval General Staff. Its deliberations were secret. Until the 1980s, very little was understood about its inner workings. Strategic planning was the General Board's primary purpose until the creation of a bona fide naval planning staff in 1915—the office of the chief of naval operations (OpNav). In the meantime, the General Board had developed into the principal authority for approving ship designs and fleet force structure—the quantities of ships and in what ratios to each other—and then ensuring these aligned with existing national strategic requirements. In many ways, its heyday was the period of naval arms limitation after the signing of the Washington treaties in 1922. The Board, well known in executive political and military circles until 1945, was virtually forgotten by history after its disestablishment shortly after the creation of the Department of Defense.

If any story qualifies as untold it was the role the executive planners of the General Board played in the building the "treaty fleet" during the period between World Wars I and II. Instead of a group of reactionary senior officers opposing naval limitation and innovation at every turn, I found instead a very dynamic, open-minded organization worried about how to execute and defend American policy abroad. The General Board served as the nexus for arms limitation preparation and implementation and absolutely set the agenda for naval construction programs. The Board was not small group of innovators and reformers operating "outside the system." Rather, it was central to the Navy's strategic processes. There was a broad tolerance within the organization for new and even radical ideas. Men like William Sims, William V. Pratt, Richmond K. Turner, William Moffett and others, who have often been regarded as mavericks or lone agents, were integral to, and comfortable with, the processes of the General Board. Admirals like Mark Bristol, Frank Schofield, and especially Hilary Jones (who was accused of seeing the "the world through a porthole") were in fact far more open and circumspect in their tolerance for new ideas than has been previously presented.[2] A more accurate description of the impact of the General Board is that the fleet that fought the Japanese to a standstill in 1942 was very much the creation of the Board and the compromises it had made during the "treaty" era.

The General Board, and by extension the U.S. Navy, was forced to consider how to project power in the far reaches of the Pacific without secure land bases for shore-based logistics. This in turn led to the development of a measurably different fleet than would otherwise have been built, a fleet that ironically was more suited to the vast reaches of the Pacific because it could operate nearly autonomously from the sea. This new conception of sea power was reflected by the treaty-built fleet in a variety of innovative modernization programs and initiatives. These included advanced mobile bases (to include immense floating dry docks), ubiquitous embarked naval aviation, and long radius of action surface ships and submarines as the principal elements. The fortification clause of the Washington Treaty was the unintentional father—and perhaps the General Board the midwife—of the modern power projection fleet, especially its sea-basing component, that is the critical core of the U.S. Navy of the early twenty-first century.

Acronyms

The U.S. Navy may be among the first of twentieth-century military organizations to inflict that bane of modern military verbiage, acronyms, upon the military culture of the country and later the world. As this study shows, the Navy was well advanced into the use of its own peculiar system of acronyms at a very early point in the century. In part this was due to its need to maintain an absolute minimum of verbiage in communications, first due to limited space on masts for signal flags, the requirement for brevity with signal lights, and the continuation of this policy for radio-communications.

AA	Anti-aircraft, often refers to anti-aircraft guns and batteries.
AF	U.S. Asiatic Fleet
ABSD	Advanced Base Sectional Dock
ARD	Auxiliary (floating) Repair Dock
BB	battleship
BuAer	Bureau of Aeronautics (U.S. Navy)
BuOrd	Bureau of Ordnance ("the gun club")
BuEng	Bureau of Engineering
BuNav	Bureau of Navigation
BuC&R	Bureau of Construction and Repair
BuMed	Bureau of Medicine and Surgery
BuShips	Created in 1940 by the consolidation of BuEng with BuC&R
CA	Heavy cruiser, originally Armored cruiser
Capt.	Captain. In the Navy this rank equal that of Colonel in the Army
CB	battlecruiser
Cdr.	Commander. In the Navy this rank equals that of Lieutenant Colonel in the Army
CF	Flight Deck Cruiser (also designated CLV)
CINCAF	Commander-in-Chief U.S. Asiatic Fleet
CL	Light cruiser

CNO	chief of naval operations
CV	Aircraft carrier
CVL	Light aircraft carrier
ENS	Ensign. In the Navy this rank equals that of Second Lieutenant in the Army
GB	General Board of the United States Navy
GPO	Government Printing Office
IJN	Imperial Japanese Navy
kts	knots
Lt. Cdr.	Lieutenant Commander. In the Navy this rank equals that of Major in the Army
Lt.	Lieutenant. In the Navy this rank equals that of Captain in the Army
Lt. (jg)	Lieutenant, junior grade. In the Navy this rank equals that of First Lieutenant in the Army
NARA	National Archives and Records Administration
NHC	Naval Historical Center
nm	Nautical Mile
NWC	Naval War College
NTS	Naval Transportation Service
OpNav	Office of the Chief of Naval Operations
PBY	Refers to the Consolidated PBY Catalina patrol aircraft/seaplane
PHGB	Proceedings and Hearings of the General Board
Rear Adm.	Rear Admiral. The highest (and) permanent flag rank in the interwar Navy. Depending on seniority, this rank equals Brigadier General to full General in the Army. The Navy did make vice and full admirals for its top two officer ranks, but only temporarily and only for force and fleet commanders. This custom changed with the advent of World War II.
USN	United States Navy
XO	This designation used in front of ship class indicates a combatant type converted from another type of ships, usually a commercial ship, e.g. XOCV is a converted aircraft carrier.

Chapter 1

Introduction: The Navy, Treaties, and Innovation

The focus of this book is the U.S Navy during the period of the "treaty navy" from 1920 to 1937. This nickname derived from restrictions that resulted from the naval arms limitation treaties signed in Washington in 1922 and in London in 1930 and 1936. Most of the literature on the "treaty navy" addresses proximate effects of the interwar treaty system on the Navy including naval innovation.[1] "Treaty system" refers to the naval arms limitation system inaugurated in Washington in 1922. The treaty system effectively collapsed with Japan's final withdrawal became permanent in January 1937.

The literature that addresses the profound impact of the fortification clause (Article XIX) of the Washington Naval Treaty on design and innovation of the interwar U.S. Fleet is scant. This clause established the status quo for fortifications in the Pacific and effectively prohibited the United States from developing new naval bases in the western Pacific or to augment its only developed base in the Philippines. The fortification clause was a fundamental, root cause that channeled innovation in the interwar Navy. There are three much-neglected elements relating to how the treaty system influenced innovation in the interwar Navy: the General Board of the Navy, the fortification clause, and the details of the construction programs that resulted in the "treaty" fleet.

First, the General Board played the critical organizational role in linking the treaty system with innovation in the design of the fleet. Particularly astonishing, given the hierarchical nature of the U.S. Navy, was the General Board's tolerant and consensus-driven process that led to an environment highly favorable to creativity and innovation. Additionally, the restrictions on the Navy's ability to build or improve its western Pacific bases caused the Navy to develop and emphasize a new approach to overseas logistics support and power projection ship designs due to the inadequacy of suitable basing ashore. Finally, the Navy's efforts in developing a wide range of programs and ship designs went well beyond the often simplistic battleship-versus-aircraft carrier dichotomy that dominates the literature about U.S. naval innovation during the period. Because of these factors, the fortification clause substantially contributed to the transformation of the U.S. Navy over the period from a "base-bound" fleet to a base-independent fleet.

There are two prevailing viewpoints that hold sway in describing the environment that enabled these innovations to take place within the Navy in the period between the World Wars. The first view identifies the battleship, the doctrines of Mahan, or a combination of both as predominant—and negative—influences on how the Navy viewed the world and prepared the fleet for war during the period. This view, for the most part, paints the U.S. Navy as aristocratic, conservative, and hostile to change while at the same time allowing for a few mavericks who developed their ideas despite "the system." Success in World War II, based on this view, followed from the crippling of the battleship line at Pearl Harbor, the serendipity of the survival of the Pacific Fleet aircraft carriers, and the industrial might of the United States, which—given time—would have prevailed in any case against the overmatched Japanese.[2]

In contrast, there exists a more recent and nuanced view of the interwar Navy as an innovative institution. This view examines the Navy from organizational and bureaucratic perspectives. Additionally, the nuanced view identifies and sympathizes with the considerable problems the Navy and its political masters found themselves facing after World War I. These problems included the Anglo-Japanese Naval Alliance, the unstable situation in China, an ongoing naval arms race between the victors of the late war, ongoing tension and war scares vis-à-vis Japan, and, finally, an American public and Congress tired of the cost of war and military hardware—especially expensive capital ships.

In November 1921, U.S. Secretary of State Charles Evans Hughes, at the behest of President Harding, invited the major naval powers to Washington, D.C., for a conference on the "limitation of naval armaments" in response to the postwar challenges.[3] The Navy found some of its problems solved: the Anglo-Japanese Alliance was not renewed and it was now a matter of international agreement that the U.S. Navy was "second to none" because the Washington Naval Treaty codified parity between the U.S. and Royal navies. It also established the U.S. Navy's superiority in capital ships over the Imperial Japanese Navy.[4] Simultaneously, the agreements drastically reduced projected naval expenditures. These benefits came at a cost. The price was the virtual elimination of the foundation for any future American defensive strategy in the western Pacific—advanced bases for the theoretically superior American Fleet—because of the fortification clause of the Naval Treaty. However, this cost was considered minimal given that several other political treaties were also signed at Washington, the so-called Nine Power and Four

Power Pacts. These treaties seemed to guarantee the long-term prospects for peace and stability in the region.

The nuanced view of the Navy is based on the problem of defense of the Philippines and its solution. The Navy was concerned with solving the problem of how to fight in distant waters without bases—in particular, how to project naval power over a distance of almost ten thousand nautical miles to defend America's Asian possessions. This very Mahanian conundrum had its basis in the reality of logistics, especially in the modern age where ships rely on fuel for their propulsion. Advocates of the nuanced view argue that the development of naval aviation, built around large, fast, long-range aircraft carriers was no accident. They further argue that the emergence of the aircraft carrier as the dominant weapon system/operational concept in the Pacific War was more due to foresight and planning than to Pearl Harbor, serendipity, and the contingencies of war.[5]

This book builds on the second view, but with the focus on the fortification clause of the Washington Naval Treaty (sometimes known as the Five Power Pact or Treaty). It examines the problem the Navy perceived the fortification clause posed. This problem framed the strategic context within which innovation occurred in the U.S. Navy of this period—the lack of suitable bases to support the Navy's anticipated strategy in the event of war with Japan in the western Pacific. Not only did the Navy attempt to solve this problem by trying to "get around" or creatively interpret the treaty, but in a more profound sense, its parallel efforts to approximate or virtually create the capability that forward bases give a Navy—without the actual bases ashore—makes for a fascinating study in innovation. These efforts, made amidst the constraints and ongoing frustrations of more naval reductions, led the Navy to fundamentally change its ethos.

The U.S. Navy went from having a "land-based" mindset to a power projection, sea-based mindset. The Navy expanded with war and massive industrial resources along these conceptual lines into the most land-independent navy history has ever witnessed: the massive fleet of World War II. The legacy of this fleet is still a centerpiece of U.S. maritime strategy today.[6]

Innovation

Another theme of this book is military innovation. Land warfare already offers an instructive case given the limiting clauses of the Treaty of Versailles and their impact on German military innovation in the interwar period. The absence of material resources clearly contributed in a major, unexpected, and positive way to German military innovation during the interwar period. General Hans Von Seeckt is often credited with orchestrating the critical organizational, institutional, and doctrinal reforms that later earned the sobriquet of Blitzkrieg, a type of mechanized warfare best typified by the first German land campaigns of World War II.[7] In many ways the U.S. Navy and Marine Corps, convinced that they too were starved for resources, developed a unified doctrine of maritime blitzkrieg, although perhaps not as consciously as their German counterparts. In both cases, treaty constraints were the underlying cause of resource constraints.

Parallels between the German case on land and the material reduction in naval armaments during this period are striking. They suggest that one unintended effect of naval arms limitation—at Versailles, Washington, Geneva, and London—was an increase in various types of innovation (organizational, institutional, tactical, and technological) by the major naval powers involved (Japan, the United States, and the United Kingdom). Nations innovated in order to make up for the perceived material deficiencies imposed by the various treaties.[8] The naval treaties were constantly on the minds of the Navy leadership of the interwar period, but none more so than the General Board of the Navy. Examination of the General Board records for the period reveals a clear linkage between the treaties and innovation. The treaties, especially those signed in Washington in 1922, established strategic and resource limitations within which the Navy had to operate. Moreover, the General Board was the locus where treaty preparation and implementation, building policy, and war planning all intersected.

The point of departure for this analysis centers on the existence of a problem: the fortification clause of the Washington Naval Treaty. In his book *Military Innovation in the Interwar Period*, author and professor emeritus of history at Ohio State University Williamson Murray discusses factors relating to military innovation, arguing that "specificity," which he defines as "the presence of specific military problems the solution of which offered significant advantages to furthering the achievement of national strategy," has been present in "virtually" every case of successful innovation.[9] Using Murray's definition of specificity, this study focuses on the issue of the fortification clause and solutions to ameliorate its impact as a basis for a broad analysis of the interwar U.S. Navy. Did the Navy see itself as having a problem? If so, what solutions existed? How many of these solutions were attempted? What did they look like? Were any solutions rejected?

Norman Friedman, Thomas C. Hone, and Mark D. Mandeles coined the term "levels of analysis" in their discussion at the beginning of their book *British and American Aircraft Carrier Development, 1919–1941*. This approach combines social science and historical methods by looking at the interactions and dynamics of organizations and sub-organizations as well as individuals within institutions. Key to this approach are definitions centered on "rules" and "players." According to Friedman et al., institutions are "society's rules of the game": "Institutions establish a stable and predictable pattern to human interaction. The term 'institution,' therefore includes formal contracts between individuals, informal codes of conduct, conventions . . . regulations, laws, charters and constitutions."

Likewise, the "players in the game" include individuals and organizations. Friedman et al. define organizations as "a group of individuals who follow particular, often formal, rules in their dealings with one another and outsiders."[10] In short, "levels of analysis" refers to analyses along several lines of inquiry focusing on how people interact and accomplish tasks as individuals, within organizations, and in tandem with other organizations within institutions. For example, Adm. John Tower represents the individual level, the Bureau of Aeronautics (BuAer) the organizational, and the U.S. Navy corresponds to the institutional level. Implicit in this approach is that social groupings and their interactions are important, but so are their constituent parts, especially those individuals in leadership and management positions.

A second approach to analyzing innovation is suggested by Alan Beyerchen, a history professor at Ohio State. Beyerchen also uses a "levels" methodology in discussing innovation during the interwar period. His approach draws on the levels of war—tactical, operational, and strategic—to analyze radar development during the period. He calls his "distinctions . . . arbitrary and overlapping—yet as useful—as the analogous distinctions among tactics, operations, and strategy in military theory." Table 1 below shows these distinctions.

Beyerchen's principal example illustrating this approach comes from the British development of radar for air defense by Sir Hugh Dowding and others prior to the Battle of Britain in the fall of 1940. In this particular case, radar was the equipment, fighter direction and radar intercept the operational concepts, and the entire air defense infrastructure (including training) was the technological context equating to the strategic level of war.[11]

Another important area to keep in mind is the political and diplomatic context for innovation during this period. The naval treaties should be viewed as a dynamic process, a systemic continuum and not as stand-alone events. The system inaugurated in Washington in 1922 was a promising attempt to implement an arms limitation system that would contribute to sustaining the peace over the long term. This dynamic system intersected and interacted with the organizational dynamics within the U.S. Navy. Moreover, this treaty *system* influenced the Navy in a variety of areas—policy, building, and innovation—on a day-to-day basis. After the

TABLE 1. Innovation Relationships

Context	Technological change	Strategy
Procedures	Operational change	Operations
Equipment	Technical change	Tactics

ratification of the Washington treaties in 1922, there was always another conference or preparatory meeting for another conference in the works. The intent was to continue the process of arms limitation and control, which diplomats and politicians hoped would lead to more stability and more arms reductions.[12] These arms conferences have often been treated by various writers as isolated events from each other and from naval and military policy-making. The various interwar treaty conferences and agreements were not isolated diplomatic events. They were linked. Additionally, discussion of military innovation during this period has often been secondary or ignored in the works that address arms limitation at that time.[13] The treaties were more than just a set of rules; they established the framework for an ongoing process and the critical strategic environment within which innovation occurred.

Finally, this book also employs Thomas Kuhn's language on paradigms and institutional thinking as a means to address how the fortification clause may have influenced the thinking of naval officers during the period. Kuhn's approach provides a means to understand innovative processes within institutional and cultural paradigms. In particular, the book addresses the fortification clause not just as a strategic problem, but as a possible anomaly to the Navy's overall conception—or paradigm—of sea power.[14]

Structure of the Book

Chapter 2 examines the General Board of the Navy. The General Board was the key organizational entity charged with the implementation of the naval limitation treaties as they applied to the Navy's building policy. Chapter 3 addresses the genesis of the strategic problem posed by the fortification clause for the Navy. This problem had its roots in the Navy's institutional conception of sea power as articulated by A. T. Mahan. This chapter defines the concept of sea power as a paradigm and how it was essential in shaping the Navy's institutional attitude in response to the fortification clause. Chapter 4 narrates the interaction between the treaty system and the General Board until 1937. It highlights how this interaction influenced strategic decision making and force structure as the Navy built a "treaty" fleet. In 1921 and 1922 the Navy chose to create a "balanced fleet"—balanced in the sense that theorist Sir Julian Corbett meant, entailing a multi-mission, and therefore a multi-platform, Navy.[15] The Navy did this for a variety of reasons, the chief one being that the Navy wasn't allowed to construct any new battleships or battle cruisers for at least ten years but was still determined to construct all the other things allowed it by treaty.

Chapters 5, 6, and 7 focus on three areas: battleship modernization, the development of naval aviation solutions such as the flying-deck cruiser (which was never built), and the interwar mobile base project. All three areas were direct outgrowths of the Washington Conference, germane to and influenced by the ongoing naval conferences and agreements of the period.[16] These three areas also align, respectively, with the tactical (battleship), operational (naval aviation), and strategic levels of war (mobile bases).

Chapter 8 provides a comparison with the experience of the other navies of the period. This discussion includes Germany and the impact of the naval limitations of Versailles. Naval innovation in Germany is addressed given the extreme naval limitations placed on them due to Versailles and despite a severe lack of resources. This will serve to illuminate further the paucity of logistic support occasioned by the fortification clause for the U.S. Navy. This comparison will also encompass the two other chief naval powers at Washington—Great Britain and Japan—and examine how treaty constraints and strategic context influenced them.

The final chapter reexamines the linkages between military innovation and arms limitation. It highlights insights about the factors affecting innovation in the U.S. Navy during the period and suggests some unintended consequences that resulted from the treaty system. The discussion closes by examining the relationship between innovation and fiscal, strategic, and political constraints during this period.

Chapter 2

The General Board

The General Board had a limited scope of activity and operated under the predominant influence of senior officers on the verge of retirement who were out of touch with new weapons and tactics.
Waldo H. Heinrichs Jr., 1973

Much maligned during its lifetime, the General Board was perceived by many USN officers . . . as a kind of Star Chamber. Its deliberations were secret, and, as the senior advisory board to the secretary of the Navy, it had the last word on ship design, and its recommendations were always influential in the areas of arms control policy and naval strategy. Because it worked secretly . . . its recommendations were often seen as the result of an arbitrary process.
Norman Friedman, Thomas C. Hone, Mark D. Mandeles, 1999

The General Board of the Navy was arguably the United States' first "general staff." The U.S. Army did not have its own general staff until after the Root Reforms in 1903.[1] The General Board is often described as merely an "advisory body" to the Secretary of the Navy, however, its real functions covered all matters of policy and strategy pertinent to the Navy from 1900 to 1950. This unique organization influenced innovation because of its balanced membership and its organizational function as the nexus where policy was translated into force structure. Its members were senior and mid-grade officers of proven experience and promise. These officers contributed to the unique character of the Board in their openness to experimentation, their collaborative organizational approach, and flexibility. The treaty system forced the Navy and the General Board to focus on solving problems (such as the fortification clause) with a variety of ship classes instead building an overpowering fleet of capital ships (battleships). Under the Washington Treaty, navies could build everything except capital ships. At least two of the ship classes allowed by the Washington Conference were radically different from those of the previous century—the aircraft carrier and the submarine. The General Board also prepared policy recommendations for upcoming arms conferences.

The structure, development, and character of the General Board determined its role in naval innovation. Because of these organizational attributes the members of the Board often served as advocates or agents of innovation. Looking at how this unique organization came to serve as an organizational nexus and then examining the elements of how it was structured and functioned during the interwar period will further illuminate how its members influenced innovation. The General Board had the characteristic of promoting collaborative solutions to problems. The treaty system, especially the Washington Naval Treaty, prompted Navy officers in general—and the General Board's members in particular—to solve problems more collaboratively.

The Board did not operate as an isolated organization. Members brought, as is only natural, their pre-existing organizational biases and professional relationships to the mix. The effectiveness of the General Board depended on a collaborative approach, which in turn grew from the way the Navy's Officer culture operated during the interwar period. Part of the explanation for why the Navy Officer Corps was collaborative can be found in the structure and size of the officer corps at the advent of the interwar period. In 1922, the U.S. Navy consisted of 7,855 officers and 89,482 enlisted sailors.[2] Promotion for the officers depended on performance at sea and was typically slow during a time of peace. By the time an officer had reached the rank of Lieutenant Commander (Lt. Cdr)—the rank at which he might first expect to attend the Naval War College or be assigned to the important staff jobs in Washington (including the General Board)—he had the expectation of either personally knowing every other officer of the same rank or was at most one or two acquaintances removed from knowing one. He also probably had a mentor, usually a former commanding officer, who looked after his interests. This small community, which only became more intimate as one climbed the ladder of promotion, was close-knit and familiar to a very high degree. Almost to a man the interwar Navy officers were graduates of the Naval Academy. They had been working for and with each other for a very long period and their tolerance for "mavericks" was very high, especially if these mavericks continued to get promoted. Four admirals, William Sims, Bradley Fiske, William Moffett, and even William V. Pratt, could be categorized as reformers or even mavericks. However, they managed to work within the system instead of against it. It seems their peers and seniors never questioned that most important of Navy values—their loyalty to the service.[3] It was within this context and from this population that officers were selected for the General Board of the Navy.

The General Board investigated and recommended a range of solutions and building policies to ameliorate the constraints of the Washington Naval Treaty. In addition to recommending upgrades to the battleships retained under the Treaty, they turned their attention to new or previously underappreciated ship classes and concepts such as aviation and cruisers. The result of the confluence of these factors was that it was the General Board who crafted a program to build a balanced Navy in direct response to the new strategic context conferred by the Washington Naval Treaty.

* * *

The General Board of the Navy was central to how the United States interacted with the treaty system. Its members influenced Navy policy and programs in a way that was meant to wring every advantage possible allowed under the Washington Naval Treaty and the later London Naval Treaty of 1930. Board members crafted advice that was also meant to shape future treaty negotiations to the advantage of the United States. They saw themselves as a group of subject matter experts on the clauses of the naval treaties. Many historians see the General Board as an archaic and increasingly irrelevant advisory body to the Secretary of the Navy, though.[4] Although some recent naval histories use General Board–derived sources to develop their narratives and arguments, the General Board remains a misunderstood organization.[5] Technically, the Board's statutory role was merely advisory. However, its actual influence was much greater, especially during the interwar period.

An example of how the Board was sometimes misunderstood by the administrations for which it worked is illustrated by the arrival of the Hoover administration in 1929. Hoover had ambitious plans that included not just naval limitation but naval disarmament.[6] Hoover had been a special delegate to the first Washington Conference and had supported a much more comprehensive plan of disarmament, including the abolition of submarines and air forces. Now that he was president he intended to make disarmament one of the hallmarks of his administration because he believed disarmament would provide for a lasting peace. Soon after his election, Hoover called on the members of the Board to explain the creation of their organization and to describe its role in naval policy-making. Much of the organizational history that follows comes from a memorandum submitted by the Board to satisfy President Hoover's curiosity.[7]

The General Board had its genesis in the overall trends in the professionalization and organization of American society during the late nineteenth and early twentieth centuries. Reformers like Admirals Mahan, Luce, and Dewey used the example of the uneven operational readiness and performance of the Navy during the War with Spain to buttress their case for a naval general staff.[8] President McKinley's secretary of the Navy, John D. Long, opposed any diminution of his authority as the secretary of the Navy. However, Long had been impressed by the performance of an *ad hoc* "war strategy board" (Theodore Roosevelt and Mahan were members) that met during the Spanish War.[9] Bowing under the pressure generated by the recent war, Long promulgated General Order No. 544 on March 13, 1900 and established a General Board of the Navy. Nonetheless, Long worried that the establishment of a naval general staff would create a direct line from the chief of the general staff to the president.[10] To avoid this problem, Long restricted the statutory powers of the new planning organization. In particular, the head of the new organization, Admiral Dewey, had no authority over either the administrative bureaus or the fleet. The Board simply drafted advice on a range of topics provided by the secretary.

The original membership consisted of the admiral of the Navy (Dewey), the chief of the bureau of navigation, the chief intelligence officer and his principal assistant, the president of the Naval War College and his principal assistant, and three other officers above the rank of lieutenant commander. The 1929 memorandum stated that "The purpose of the Board, as stated in the order [No.544], was 'to insure *efficient* preparation of the fleet in case of war and for the naval defense of the coast.'" (emphasis mine)[11] As such, the General Board reflected the military reform of that era in the naval sphere. Still, the establishment of the General Board was an innovative organizational initiative, meant to increase the efficiency of the Navy as a whole as well as to provide civilian administrations with formal policy and planning advice.[12] Over time, the General Board's advice became more and more influential. By the time of the Washington Conference, advice from the General Board on a particular topic was considered the institutional "party line" for the entire Navy.[13]

The Navy bureaus of that day also warrant explanation. These autonomous entities largely "ran" that part of the Navy that built ships, provided resources, and generated administrative policy prior to World War I. They worked for the secretary of the Navy and coordination between them was informal. Disputes between the bureaus were resolved at the secretarial level. The major bureaus consisted of Navigation (which included personnel management), Ordnance (sometimes called the "gun club"), Construction and Repair (C&R), Engineering, Yards and Docks, and (after World War I) Aeronautics.[14] Throughout the Navy they were known by their abbreviations: BuNav, BuEng, BuOrd, BuC&R, BuY&D, and BuAer. Each was usually headed by an unrestricted line admiral, often destined for higher command, except BuC&R which was often under the leadership of the senior construction corps officer (later the Civil Engineering Corps).[15] The General Board came to serve as a coordinating staff between the office of the secretary and the bureaus. The only people satisfied by the creation of this kind of General Board were Long and his allies in Congress.[16]

In April 1901, the Board was reorganized by another General Order (Number 43). It is not clear from the record what the impetus for this reorganization was, although the prize essay for the United States Naval Institute that year had addressed a reorganization of the Navy.[17] The principal assistants to the president of the Naval War College and chief intelligence officer (later the director of Naval Intelligence) were removed from membership on the Board. This had the effect of allowing their bosses to attend Board meetings while they acted within the bureaus in their stead. Additionally, the secretary now selected three junior officers of at least lieutenant commander rank whom he appointed to the Board. This had the effect of giving the secretary more direct representation on the Board and provided a way to develop junior officers of his choosing for further responsibility by exposing them to the weighty issues of strategy and policy discussed by the Board. More changes followed, specifically the "aid" system that added the aides (to the secretary) for operations, personnel, and material to the Board while eliminating direct representation by the bureaus. All these changes increased the influence of the secretary over the Board.[18]

A more substantial reorganization of the Navy took place with the advent of World War I and the creation, by Congress in 1915, of the independent office of the Chief of Naval Operations (CNO or OpNav). At long last the Navy had a statutory general staff in the form of OpNav. Although still not a general staff in the strict sense of that term, OpNav was not only responsible for the current operations of the fleet but also conducted contingency planning for crisis and war in its War Plans division. This act resulted in a decrease in the participation of the civilian secretary in strategic planning, but not all at once. First, the "aid" offices were abolished. Next the chief of naval operations (CNO) and the commandant of the Marine Corps became "ex-officio" (permanent) members of the General Board.[19] Thus OpNav and the General Board were organizationally linked by having the CNO or his representatives sit regularly on the Board. Meanwhile, the General Board had developed a great deal of "informal" influence and authority as the principal advisery body to the secretary on all matters of strategy, policy, and naval construction.[20] In addition to providing advice to the secretary, other Navy organizations—OpNav, the Naval War College, and the bureaus—also solicited the Board's views. The Board could, and often did, direct specific actions to be taken by the bureaus and also had its requests for action serviced by the War College and OpNav staff. Nevertheless, the establishment of OpNav provided a mechanism for the decrease in the Board's influence over time. There was now another competing organization providing military advice to the secretary (see Figure 1).

The Navy of the interwar period was a collaborative place and the General Board encouraged cooperation, and thus innovation, across the various levels of war—tactical, operational, and strategic. The Board tended to focus most on the

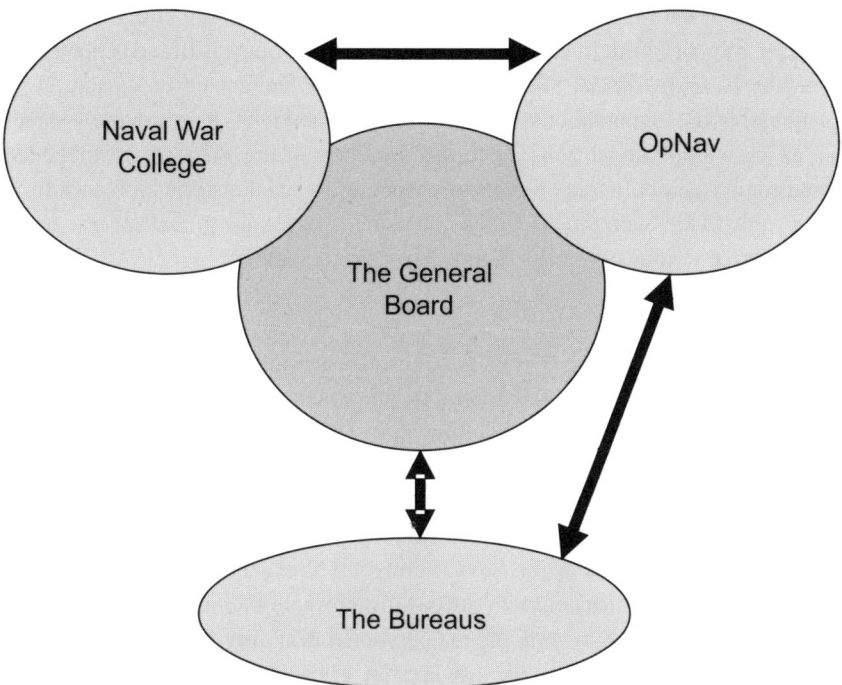

FIGURE 1. Navy Organizational Relationships during the Interwar Period. (Note: Overlap indicates membership. Lines indicate lines of communication and coordination.)

strategic level. It developed policy and applied its policy decisions to the overall design of the Navy (numbers and types of ships). OpNav's focus—inherent in its name—was the operational level of war that included planning for both current and future operations. Although its War Plans Division included a strategy "cell," OpNav focused overwhelmingly on the operational spectrum—conducting real world operations and exercises and designing plans around the fleet at hand.[21] OpNav collaborated with the Naval War College in testing and developing operational concepts using the College's war gaming process. The results of these war games were also shared with the General Board. Similarly, the War College was also linked to the General Board by statutory membership (see Figure 1). Tactical issues came up in the course of planning at OpNav, in war gaming at the War College, and in the discussions of the General Board. At the Board's hearings and meetings, technical and tactical design considerations entered the process, often through testimony by representatives of the bureaus and other experts. These meetings, especially the topical hearings, became a forum for the discussion of policy and design issues at the strategic, tactical, and operational levels of war.

By 1929, the General Board still saw itself as an important entity in the formulation of plans, policy, and strategy: "Although the General Board is not established by [Congressional] Statute, it has long been recognized in legislation by Congress. In the organization of the Navy Department it has a very definite standing as a *personal* advisory board to the Secretary of the Navy. Its membership being composed of officers of long experience and special qualification, its advice is available to the Secretary on broad questions of naval policy and specific questions referred to him from time to time." (emphasis mine)[22]

* * *

The Navy secretary and the CNO referred most issues that came across their desks having to do with naval policy, building policy, strategy, and especially arms limitation to the Board for study and comment. In 1929, the Board was composed of an executive committee as well as the "ex-officio" members—the CNO, commandant of the Marine Corps, president of the Naval War College (NWC), and director of naval intelligence—a secretary (usually a senior Navy commander), and other officers as assigned by the Navy secretary. The executive committee usually consisted of twelve officers, with seven of them being admirals either coming from operational fleet duty or going to operational fleet duty. During this time, the entire Board met regularly on the last Tuesday of every month as directed by the secretary, whereas the executive committee met at 10 AM every day Monday through Friday. Capt. Robert Ghormley, author of the 1929 memorandum that provided a history of the General Board for the Hoover administration, thought it important to emphasize at the end of his memorandum that the war plans function of the General Board had been "transferred to the War Plans division of the Office of the Chief of Naval Operations. . . ."[23]

In 1932, the Board, reflecting the influence of the CNO Adm. W. V. Pratt, was pared down by eliminating "ex-officio" membership and the executive committee became the standing membership of the Board. Pratt also requested that the Board continue to forward its advice to the CNO as well as to the Navy secretary.[24] After 1932, the Board held to a schedule of daily meetings (usually in the mornings). The Board invited testimony from a broad spectrum of experts to include non-Navy organizations such as the Army and private civilians. The informal power of the General Board remained substantial throughout the interwar period—only the outbreak of World War II and the post-war reorganization of the military made the Board seem unnecessary.[25] A measure of the Board's standing is suggested in the career of Adm. Ernest J. King (ultimately a five-star admiral in World War II). King had been a member of the General Board immediately prior to his appointment as commander in chief of the new Atlantic Fleet and shortly thereafter also became the CNO.[26]

In the hearings, studies, proceedings, and deliberations of the General Board one finds a confluence of a variety of functions. The Board addressed naval arms limitation, strategy, the recommendations of the bureaus, policy-making, Naval War College war gaming results and studies, fleet queries and feedback, and requests from the CNO for opinions. Often the Board would invite a particular organization, for example OpNav's War Plans division, to provide testimony on a specific issue. Testimony, both written and oral, was also gathered from organizations and individuals outside of the Navy.[27] The Board addressed all the available information when fulfilling its responsibility to advise the secretary of the Navy, as specified in Navy regulations. The Board drafted advice for both military and civilian policy-makers. It cast a broad net when addressing a particular issue as it drafted its recommendations. Seeking a wide range of testimony created opportunities for innovation because it tended to expose the Board's membership to new ideas from both inside and outside the Navy.

Many of the key Navy leaders who emerged in World War II as major commanders and leaders were previous members of the General Board, for example, Ernest J. King, Richmond K. Turner, and Thomas C. Kinkaid. Adm. Thomas Hart, the commander of the Asiatic Fleet at the outbreak of war in December 1941, had been the chairman of the General Board from 1936 to 1939. Often these officers either had sat for extended periods on the Board or had served multiple times over the years at various levels. One example is Kinkaid, later commander of the Seventh Fleet under Gen. Douglas MacArthur. Kinkaid testified before the Board on behalf of BuAer in the 1920s and was later assigned to the Board as its secretary from 1930 to 1931. In a similar fashion, Adm. A. T. Long had been a member of the Board from the Office of Naval Intelligence in 1921 and later served as chairman of the executive committee of the Board in 1929. In like manner, two admirals, Mark Bristol and Harry Yarnell, served as fleet commanders and later as chairmen or appointed members of the Board. The migration of officers between the General Board, the fleet, CNO, and War College was fluid and suggests that participation in the activities of the General Board was career-enhancing as well as perspective-broadening.[28]

The General Board Process

The mechanics of the General Board's processes and products highlight many characteristics of its structure and function that impacted naval innovation. First, the General Board was authoritative in the sense that its advice was often considered "the last word" on a particular issue.[29] For example, the correspondence of the interwar period is full of references to "the opinion of the Board" or "the judgment of the Board." These opinions and judgments were reference points that the secretary of the Navy, CNO, the War College, and the bureaus would use in making decisions, initiating programs, and spending money. The General Board's membership also reflects its authority within the Navy. Members for most of the period included the CNO, commandant of the Marine Corps, director of Naval Intelligence, and president of the Naval War College. Even after Admiral Pratt's changes in 1932, these officers could, and did, provide testimony and written input to the Board.[30]

The General Board process was an "open" one. There were open lines of communication between the Board and other organizations, both internal and external to the Navy. These lines of communication were not limited to governmental entities. The Board examined a broad array of ideas and testimony, either written or in formal hearings. It could invite experts from anywhere to testify on any topic of interest. All of these characteristics favored the collaborative, and sometimes confrontational, exchange of ideas. The Board posed potential courses of action about topics ranging from naval policy (such as arms limitation) to the thickness of armor on a battleship's bridge. The Board's collaboration with other Navy organizations presented opportunities to select innovative solutions, especially since the naval treaties had curtailed the Navy's traditional approaches to its tactical, operational, and strategic problems—battleships and bases.

The evidence suggests the General Board was very good at working on multiple issues at the same time. It tended to withhold judgment and instead, subjected new or competing ideas to analysis, collaborating with other Navy organizations in the process. It often simply deferred making its recommendations until a concept or ship design had been tested at the War College and then experimented with by the fleet. Results of fleet exercises, which were usually built around "battle problems" and the College's war gaming results, were used by the General Board in its deliberations. In particular, the hearing process that was used to craft advice for the secretary of the Navy shows the General Board displaying many of these positive characteristics simultaneously.

The General Board advised the secretary of the Navy using meetings and written products. The meetings could be either closed informal discussions or formal hearings by the Board. All hearings were transcribed as a secret record. Closed meetings could include the entire membership of the Board but more often were limited to the executive committee prior to 1932. Meetings could be regularly scheduled events but were usually held as needed and conformed to no regular schedule, but also included special meetings convened by the secretary of the Navy or even the president.

The focus of both the closed meetings (for which there were no formal transcripts) and the hearings was to provide advice in the form of a written study.[31]

These studies often started out in the form of a draft study or "yellow" (because they were printed and distributed on yellow paper) that could be drafted before or after a hearing. Yellows were sometimes drafted without any formal hearings at all. For most issues—policy, tactics, arms limitation, and ship design—the Board consulted a broad variety of expertise. In 1929, the Board explained the purpose and initiation of its hearings as follows: "When the General Board has a subject before it for consideration on which the advice or recommendation of materiel bureaus, other officers of the Department, or civilian experts, is desired, a hearing is held at which these various representatives are requested to be present and present such information as they may have for consideration by the Board. These hearings are recorded, bound and kept in the General Board offices, and form an excellent set of reference papers for further use. From such hearings and personal knowledge of the members, the General Board formulates its recommendations to the Secretary."[32]

The Board could use the transcribed hearings to either draft a yellow or make changes to an existing one. Once the yellow had been revised, it was submitted as a numbered "serial" to the secretary of the Navy. The General Board assigned the numbers in chronological fashion based on when the topic was received by the Board. For example, the use of submarines in warfare was referred to the Board and assigned as serial 1182 on July 6, 1923 whereas the modernization of battleships was assigned as serial 1195 on October 15, 1923. These numbered serials were retained for reference. Sometimes serials were referred to other Navy organizations (typically the bureaus) for work, or in the Navy terminology of the time "for action." Serials were also known as General Board "studies."[33]

General Order 544 stipulated that the secretary determined the agenda of the Board. In reality, the determination of the agenda was usually an interactive process. Often the CNO, as well as organizations outside the Navy such as the Army–Navy Joint Board or the office of the Secretary of State, would stimulate a dialogue that resulted in a General Board hearing.[34] The Board also considered topics forwarded to it by the fleet and the bureaus and referred advice on them up the chain of command to the Navy secretary.[35] The Board could also recommend hearings on issues or tangents that arose from the free-flowing discussions during its hearings. Sometimes, as happened frequently during the Hoover administration, topics were transmitted by the president through the Navy secretary directly to the Board.[36]

The hearing process was informal, but it did have a consistent and clear structure within which an open and honest dialogue about the issues forwarded to it could be discussed. A typical meeting would have a set issue that it was assigned to deal with. For example, a hearing was convened on March 3, 1922 in Washington D.C. to discuss aviation policy within the larger naval policy being crafted as a result of the Washington Naval Treaty. This meeting, which was the critical meeting that implemented the terms of the Washington Treaty by the institutional Navy, illustrates the range of participation, the influence of the treaty system, and the organizational dynamics of a General Board hearing.[37] It will be discussed in more detail in the next chapter, but for the purpose of understanding the general attributes of a hearing it was typical. The attendees at this meeting included the executive committee of the General Board. On the executive committee was the senior member, Adm. William Rodgers in this case. However, on nearly every occasion, the same procedure was used. The senior member, after some opening comments, would normally defer to one of the more junior admirals to facilitate the hearing. Sometimes the meeting would be facilitated by one of the senior captains on the executive committee.

Once the senior member turned the meeting over to the facilitator, the hearing was in his hands. Sometimes the Board had an initial series of questions it wanted answered and at other times the subject matter experts had prepared comments that they would make, which would then be entered into the record. Again, subject matter experts could run the gamut from civilian scientists to the heads of the various Navy bureaus. The facilitator usually asked the questions, but it was clear that any member could ask a question at any time. Sometimes the facilitator would allow the questions, and sometimes defer them for later discussion. During the entire process, the senior member often reserved his comments and rarely entered into the give and take of the hearing unless asked. Once the facilitator felt the topic had been adequately addressed, and the Board's questions answered, he would turn the proceeding over to the senior member for final comments and adjournment. The astonishing attributes of these hearings were their flexibility, cordiality, and informality.

The Board also gathered testimony from outside the Navy. The hearing process allowed the Board to consult experts on topics about which the Board members had limited experience. Gen. Billy Mitchell of the Army Air Corps testified on air power during the 1919 hearing on the future of aviation in the Navy. In this case, the scope of the testimony allowed by the Board was misinterpreted by Mitchell. He concluded that the Board agreed with many of his ideas about air power because it had politely listened to and questioned him in a non-judgmental manner. However, the Board was unconvinced by much of what Mitchell said, but listened to and explored his ideas nonetheless. This is evidence of an organization that was willing to entertain—and even support on occasion—provocative ideas as long as they contributed to progress and efficiency. But new, unproven ideas could just as likely remain in the hearing transcripts, undeveloped and unexplored, as the Mitchell example highlights.[38]

Once a topic had been explored via the hearing process, the next step was to draft a serial for the secretary of the Navy based on the testimony. Sometimes, as was the case for the March 1922 hearing, the Board already had a template—the naval policy "blueprint"—and so merely needed to amend the verbiage that already existed. The hearings helped the General Board clarify and examine as much of the available evidence and expertise as possible before referring the serial to the secretary. Serials were not always the result of a hearing and could also be independently generated by the executive committee of the Board. The clerical work was typically overseen by the Board's secretary and involved producing the yellow if one did not already exist. Once the yellow was complete (either before or after a hearing and sometimes without one) it was distributed to the other members for review. Sometimes the yellows were referred back to the bureaus for comment prior to the formal submission of the serial to the Navy secretary. In the case of the Naval Air Policy, a lengthy period of time passed after the March hearing due to extensive communications between Adm. William Moffett (BuAer), the Board, the CNO, and the Navy secretary. For example, Moffett sent a confidential memorandum directly to the Navy secretary (via the CNO) on August 15, 1922 expressing concern over the ability of the Navy to expand its naval aviation manning in time of war. This memo was in turn forwarded by the CNO, who was a member of the Board, to the General Board "as a basis for discussion in the formulation of a Naval Aeronautic Policy." The Board, in turn, addressed Moffett's concerns as it drafted the final policy recommendations in its finished report.[39]

Board secretaries, often junior officers for much of the interwar period, tended to have a great deal of latitude in drafting the yellows and even the final serials. They were subject to scrutiny because the transcripts of the hearings could always be consulted by the other members. The final serial was always signed, and one assumes closely proofread, by the senior member present on the Board, usually the chairman of the executive committee. Although no minority reports were officially made, the endorsements and yellows both provide ample evidence of disagreements and differences of opinion. However, the final serial, especially the recommendations, were the last word on the General Board's advice on any particular topic.[40] This advice was reached by a majority consensus among the members, although it is not clear what the specific rules for consensus may have been beyond a simple majority. Based on the signature of the senior member of the Board on the final serials, it appears that the senior member made the determination of consensus based on everyone's inputs, often written as marginalia on the serials.[41] This process made compromise almost a necessity. These compromises were often reflected in the give and take of hearings such as those that discussed ship design. In these, members frequently noted that a final design would have to reflect a balance between competing factors such as speed, armor, and gun size.[42] Similarly, hearings and studies for arms limitation policy involved, by their very nature, the specter of clear compromise that would be required in making recommendations for policy positions to the secretary.[43] General Board members were no strangers to compromise, although during the interwar period there were many occasions where they appeared uncompromising and stubborn. Lost, perhaps, in making this judgment are the internal compromises that members undoubtedly had to make in the writing of the consensus opinions in the formal serials. These dynamics will be more fully discussed in the next chapter.

The hearings and the drafting of the serials were interactive processes. The approved air policy discussed on March 3, 1922 was promulgated before the serials that prompted it had been formalized and approved. General Board processes were a continuum. The Board was already planning modifications to its general policy for 1923 and building policy for 1924 in late 1922.[44] Recommendations in the final serial were meant to serve as the basis for future studies addressing naval policy and ship construction as well as to serve as a record of the reasoning behind the Board's advice. If any members of the Board later had a question about the testimony on a particular topic they could consult the record to refresh their memories.

The final step was formal approval of the completed serial by the secretary of the Navy. GB 449 Serial 1140 entitled "U.S. Naval Policy" was formally published, including Moffett's August 1922 memo and other endorsements, on November 18, 1922. This serial is not to be confused with the Naval Air Policy for that year, which had been approved by the Navy secretary with the General Board's concurrence on May 16, 1922. The serial encompassed the entirety of the "U.S. Navy Policy" drafted to conform to the Washington Naval Treaty (see Appendix 2). Finally, the secretary approved the published Serial 1140 on November 23, 1922. The serials were not for public consumption but were secret to serve as reference material to document the Board's advice and to serve as the basis for future work.[45]

An optional step existed in the serial process. Occasionally the secretary would return a serial for reconsideration by the Board. This step was not taken often and reflected potential policy disagreements between the Board and the civilian administration. An example came during preparations for "a future conference for further limitation of armament" in 1924 (Serial 1239). Originally the Board had started the work at the behest of acting Sec. of the Navy Theodore Roosevelt Jr. The Board took up the issue as it prepared its serial and noted in October 1924 that "action is indefinitely postponed." Curtis Wilbur, the new secretary of the Navy, "orally" instructed the Board in May 1925 to develop an agenda for a future conference, especially to include the issue of deferring the replacement of capital ships. One proposal being considered by the governments of Great Britain and the United States was an extension of the capital ship "holiday," a building moratorium of ten years' duration that had been approved by all the signatories at the Washington conference. The Board opposed any extension of the existing "holiday" on capital ship construction and had recommended that replacement construction for this type occur as soon as permissible by the Washington Naval Treaty. In June, the secretary's guidance was again taken up and the executive committee "agreed upon a reply" after 45 minutes of discussion. There was no hearing on that date to offer additional insight into what was actually discussed. In its written response, submitted formally on June 3, 1925, the Board emphatically rejected the secretary's "tentative suggestions." Secretary Wilbur pondered this for about a month and on June 29 responded curtly, "Please reconsider this recommendation." The entire correspondence was classified "secret." The serial containing this advice was never formally approved, probably because the Navy secretary realized that the Board's recommendation was counter to the administration's policy of advocating an extension of the building holiday but did not want to go on the record as taking a position at odds with the recommendation of the General Board.

* * *

In summary, the General Board had several characteristics that influenced innovation. These characteristics often overlapped and supported each other. First among these was the Board's consensual approach to decision making that relied on compromise and that relied on, and sought out, a wide variety of opinion and knowledge on any particular issue before it. The General Board's membership was structured in a way that favored a multi-branch, "multi-ship-type" approach to force structure and strategic policy. Everyone—submarine officers, aviators, battleship sailors, engineering officers, and so on—had a seat at the table. The General Board provided the forum for the interaction of these officer communities.[46]

In addition to consensus and structure, the organizational location and authority of the General Board within the overall hierarchy of the Navy tended to reinforce the influence of its decisions across the Navy. This role conferred upon the General Board a weighty responsibility which it took seriously. Adm. William V. Pratt once referred to the General Board as the "balance wheel" of the Navy. Pratt was referring to a wheel that could be used to regulate movement in a machine (sometimes known as a speed governor). Similarly, the General Board—in Pratt's metaphor—was central to the process of developing naval policy. Although we cannot be certain if Pratt intended the metaphor in any other way, we can imagine the General Board as a wheel at the center of the Navy, representing the secretary of the Navy, with "spokes" as lines of communication and influence radiating outward to the "rim," which might be seen as the fleet and the bureaus (see Figure 1, page 13). Additionally, the Board had to "balance" its recommendations against the realities of what was politically and fiscally achievable; otherwise its advice risked being routinely ignored as being unrealistic.[47]

Stability was conferred not only by the internal balance of different constituencies within the Board, but also externally as the Board played the role of "umpire" among the bureaus, CNO, War College, and other offices in balancing requirements with fiscal and treaty realities. Also, if the resident expertise on the Board was deemed inadequate to address a topic, the Board could augment itself using the hearing process to get an even broader scope of testimony and opinion. For example, the Board often invited the bureaus or CNO to send their representatives to a hearing. These officers—especially CNO since he was a member until 1932—could also notify the Board of their intention to attend a particular hearing. The Board was also meticulous in addressing written statements during its hearing process. The cumulative effect of the Board's "open structure" led to the consideration of a broad spectrum of testimony and ideas and information that was used in the design of ships and policy. Although a broad spectrum of opinion did not guarantee that innovative solutions would be adopted, it did guarantee innovators (like Admiral Moffett) and innovative ideas a seat at the table. This might not have been the case if the Board had had a more closed structure and decision process. The Board, in part because of the original Washington Naval Treaty, but also due to its own structure and methods, also had the characteristic of promoting parallel and even competing solutions to problems. This was in no small measure due to the uncertainty about just what the future held. As several esteemed historians have noted, the Navy hedged its bets in promoting a number of different building programs from naval aviation and battleships to submarines and cruisers.[48]

General Board members believed, as did the rest of the Navy officer corps, in the value of experimentation. Experimentation could be used to examine the value of a new concept or design. Alternatively, as in the case of the aircraft carrier, it was used to refine and improve existing concepts, ship designs, and tactics. Thus innovation could occur, and often did, in both the examination of existing technology and concepts as well as new concepts and untried (but theoretically feasible) technology. Finally, the General Board was flexible. It was flexible as to its procedures, personnel, scheduling, and prioritization of tasks. It could call a hearing on just about any topic of interest and bring in the testimony of literally anyone willing to make the trip to Washington D.C.

During the interwar period the mechanics and membership of the General Board changed with the times. The fundamental process remained one where the agenda was set principally by the secretary of the Navy. However, the Navy was a collaborative collection of organizations and individuals. The Board was at the center of these collaborative processes, especially those implementing the treaties that produced a balanced U.S. Fleet. General Board members served as agents and partners in innovation because of the balanced makeup of the General Board and its role as the nexus where policy was translated into force structure.

Chapter 3

U.S. Sea Power and the Washington Conference

Sea Power is not made of ships, or of ships and men, but of ships and men and bases far and wide. . . . Manifestly the provisions of the Treaty presented a naval problem of the first magnitude that demanded immediate solution. A new policy had to be formulated which would make the best possible use of the new conditions.
Capt. Frank H. Schofield, 1923

Navies are prone toward technological and engineering solutions to problems—modern navies especially. Going to sea, for war or commerce, has tended to reflect the limits of the technological developments in the society so inclined. The U.S. Navy of the interwar period was no exception and may have exceeded some of its peers in its infatuation with technological solutions to naval problems. Navy officers, paradoxically, tend to be of a conservative mindset. This is because solutions are often explored on the basis of technological improvements to an established or proven design and within existing doctrine. Most innovation, or work, tends to occur within an accepted institutional, cultural, or scientific paradigm. For the Navy, the paradigm of most importance was sea power and it continued to be so in the interwar period. Instead of making radical changes during this time, it tried to adjust its solutions and designs to fit an existing paradigm or conception of sea power that was rooted in the teachings of A. T. Mahan.

The design bywords in the U.S. Navy of the interwar period were stability and efficiency—and they remain so today. Stability is essential aboard a ship. Changes to a complex and dynamic system like a manned warship have unintended consequences. Changes can affect everything from the quality of life aboard ship—the loss of hot water—to a complete loss of a weapons systems or propulsion. Efficiency is related to stability. Changes that do occur are judged not only on how they affect a particular area, for example damage control, but their impact on other related systems and functions. If the net effect of a change, especially a radical change, is judged to decrease the overall efficiency of the ship, it is often not implemented, despite its technical or conceptual promise. This same sort of thinking can be extrapolated to higher systemic levels such as the design of a fleet.[1] Thus change is often met by naval cultures with healthy skepticism and is often incremental. This skepticism contributes to a more conservative frame of mind—navies are among the most hierarchical of military cultures, already predisposed by tradition and structure toward a conservative ideology.[2] Other approaches, especially untested organizational or operational ones, stand a chance of being undervalued or ignored altogether by this mindset.

This is not to say that the U.S. Navy had no history of innovating. Naval service also breeds decisiveness. When the need to change and adjust reaches the point where the ship, or the institution, is perceived to be "sinking," then change often occurs rapidly and comprehensively. When a change is given to the helm, the entire ship turns—not just the captain.[3] This phenomenon also tends to be reflected in the way navies change as institutions. The U.S. Navy had led the way among the two services in reorganizing itself in the last part of the nineteenth century. It had established the Naval War College and eventually created the General Board of the Navy, a scaled-down version of a naval general staff. These improvements occurred before the more famous Root Reforms that restructured the U.S. Army. However, by the end of World War I many of these organizational reforms were complete and there no longer seemed to be a pressing need for further change.[4]

The U.S. Navy's commitment to stability and efficiency was especially true for its battleships. This ship type had weathered millennia—only changing with improvements in technology and design—and remained the backbone of all the major modern fleets. So powerful was the idea of the battleship as the final arbiter of naval power that nations saw its limitation as a means to halt, and eventually eliminate, naval competition. The victors of World War I saw the dreadnought naval race as one of the causal factors of that ruinous war. When naval arms limitation was initiated in spectacular style at Washington in November 1921, it placed the traditional, battleship-oriented U.S. Navy on the horns of a dilemma. How could the Navy accomplish its strategic tasks with a doctrine and strategy based on the battleship and overseas bases when the Washington Treaty had severely limited both of these pillars of sea power? Navy leaders, in attempting to solve this problem, remained wedded to their traditional concept of sea power, so much so that it paradoxically resulted in innovations that might not otherwise have occurred. Had the Washington conference not been held, and its attendant naval limitations not been implemented, and had the Navy not remained committed to a Mahanian concept of sea power centered on the battleship, it probably would not have followed the path of innovation that it did during the interwar period. Similarly, success in World War II in the Pacific may not have come as early as it did.

Through an evolutionary process that was primarily influenced by the system initiated at the Washington conference, but also by powerful political and budgetary factors, a plan was fashioned to build a "treaty fleet" that could project power to the western Pacific in the event of war with Japan. The mature plan was exemplified by the General Board's 1929 building program and involved the construction of a completely different sort of fleet—one that could range for thousands of miles and apply sea power in the absence of preexisting overseas bases. This was the basic design criteria for the fleet that absorbed incredible punishment at the hands of the Imperial Japanese Navy during the first year of the Pacific War and still managed to fight it to a standstill by late 1942 without any of its prewar battleships.[5]

In sum, U.S. naval strategy revolved around the successful application of sea power, which was universally understood by Navy officers to be comprised of a fleet, domestic and overseas bases, and a robust merchant marine. If any of these three elements was inferior to that of an adversary it must be compensated for by an increase or adjustment in one of the others. It was for this reason that the Navy came to support the ratio system established by the Washington Naval Conference. If the United States built to treaty limits it would have a fleet superior by 5–3 to the Japanese. Such a fleet compensated for the limitation on overseas basing conferred by the fortification clause.[6]

The Navy of the treaty period had views on sea power that were derived from the writings of A. T. Mahan, contrary to what some historians have written.[7] After the Washington conference, Navy officers were forced to think in terms of fighting a distant naval war in the absence of any significant bases in the projected theater of operations. This concept of sea power, and the challenge posed to it by the fortification clause (Article XIX) of the Washington Naval Treaty, determined the way in which innovation occurred in the interwar Navy.

The Washington Naval Conference

Secretary of State Charles Evans Hughes dramatically initiated the Washington Naval Arms Limitation Conference on November 12, 1921 by proposing the Harding administration's plan for a reduction in naval armaments. The main element in this proposal revolved around a fixed ratio for capital ship tonnages (battleships and battle cruisers) for the United Kingdom, the United States, Japan, France, and Italy. The allowance followed the formula 5–5–3–1.75–1.75, respectively. The other important features of Hughes' proposal included: the scrapping of over 1.8 million tons of older ships and ships under construction (with the United States scrapping the most), a "naval holiday" on capital ship construction of ten years' duration that included prohibitions on replacing existing capital ship tonnage, and a proportionate ratio to be applied to auxiliary ships.[8] Missing from Hughes' initial proposal was any hint of implementing a status quo for naval base fortifications or improvements in the western Pacific. The article that addressed fortifications came out of the subsequent negotiations during the conference.

Hughes' masterful presentation on November 12 did not lead to immediate consent to his plan by the other major powers. Significant areas of dispute remained to be resolved over the course of the next three months. Between November 12, 1921 and February 6, 1922, when the treaties were signed, the issue of limiting fortifications in the Pacific arose as a means to ameliorate Japanese sensitivity at being ranked in an "inferior" position to the United States. Some members of the Japanese delegation—but not its chief delegate, Navy Minister Baron Kato Tomosaburo—took issue with the 3–5 ratio that Japan's fleet was assigned relative to the U.S. Fleet (sometimes referred to as 6–10). The Japanese delegation had been sent to Washington with instructions to accept no less than a 7–10 ratio in capital ships vis-à-vis the United States. The Japanese position was partly attributable to their American counterparts. In the first place, the Japanese and American navies shared a common understanding of sea power based on the writings of A. T. Mahan. In the second place, the Japanese made their calculations based on their own intelligence estimates of what the Americans projected for the size of their fleet.[9]

Mahan had discussed the issue of maintaining the operational effectiveness of fleets over extended distances in his lectures on strategy. Using the Russian Navy at Tsushima as an example, he showed how the effectiveness of a fleet, despite numerical superiority, can be negated by time and distance. Mahan argued that these factors, without adequate basing or maintenance in naval yards, affect crew fatigue and the physical condition and seaworthiness of the warships, especially those powered by steam engines.[10] It was on this basis that American planners from the War Plans Division of OpNav formulated a rule of thumb holding that, "for every 1,000 miles a fleet steamed from its base, it lost 10% of it fighting efficiency."[11] Without bases beyond Hawaii, these planners estimated the U.S. Navy would need a minimum of a 10–6 ratio to maintain a slight superiority over the Japanese Fleet which would be located much closer to its own operating bases. The Japanese had used the same process to deduce that they needed at least at 7–10 ratio with respect to the U.S. Fleet in order to maintain a margin of superiority. In addition, it appears that Japanese naval intelligence had acquired key elements of the naval portions of the U.S. Orange War Plan that confirmed their calculations.[12]

The 10 percent chasm between the United States and Japanese positions, a difference that could completely derail the conference, was resolved within the Japanese delegation by Baron Kato. Overruling subordinates and building support among senior admirals in Japan like Admiral Togo, the hero of Tsushima, Kato offered the Americans a counterproposal. The Japanese would accept an inferior ratio if the Americans agreed to the status quo for their limited number of bases in the western Pacific—essentially Guam and Cavite in the Philippines.[13] This proposal was formally agreed to on December 22, 1921. Although some details remained to be ironed out, this agreement committed the United States to an inadequate system of naval bases to support its interests should war come to the Far East.[14] Of course the whole idea was that this agreement would reduce the possibility of war with Japan by lessening the tensions between the two nations associated with naval competition. However, the U.S. Navy had previously designed its ships on the basis of extensive support ashore and had hoped to measurably increase its facilities in both Guam and the Philippines in order to support its assumptions in the developing Orange War Plan. Either the Orange Plan would have to be considerably modified, or an entirely new Navy would have to be designed and built, and existing ships modified. In the end the Navy did both, but not all at once and only grudgingly modified its preference for an Orange strategy that immediately deployed the bulk of the fleet to secure the Philippines.[15]

By proposing the status quo clause, Kato managed to make the treaty politically acceptable at home—although there was still opposition from elements of the Japanese Naval general staff.[16] This concession also satisfied the mostly civilian American delegation. There were no fortifications approved or funded for the western Pacific and now there was no need to build any. In the view of President Harding's administration, the fortification clause saved the United States money. Baron Kato perhaps displayed a better understanding of the American political system than the American delegates did in balancing the ratio with the fortification clause.[17] Kato "bet on the come," gambling that the Americans would neither build to limits for what they were allowed—aircraft carriers—or extensively in those categories of ships that were not limited, such as cruisers.[18] Thus, while having potential parity with the Japanese in the western Pacific, this clause in fact worked out in favor of the Japanese if the Americans built anything less than they were allowed. As it turned out, this is precisely what happened. The Americans did not build to match the Japanese. U.S. Navy leaders, on the other hand, were acutely aware of how this affected their strategic plans. While continuing to lobby both Congress and the various administrations to "build to treaty limits," Navy leaders also looked for innovative ways to ameliorate what they saw as the decreasing strength of their fleet relative to Japan's.

Kato also knew that what was prevented in peace was not prevented in the event of war. Japan's retention of the League of Nations Mandate island groups between Hawaii and the Philippines would be directly across the U.S. Navy's line of advance in any plan to relieve the Philippines or move the fleet permanently to the western Pacific. In the upshot, Kato worked very hard in crafting the final form of the article in the face of opposition by factions in the Army and the government. These factions opposed the inclusion of some of Japan's southern island groups (Formosa and the Pescadores) in the nonfortification status quo. However, by the end of January 1922 Kato's personal efforts were rewarded by his government's assent to the signing of the treaties.[19]

Another dispute arose over Secretary Hughes' proposed extension of limitation ratios to auxiliary classes of ships. This dispute had its origins in the capital ship ratio. The French delegation, outraged by Hughes' proposals, demanded a larger ratio than Japan. However, Secretary Hughes appealed to the French Prime Minister Aristide Briand who then directed his delegation to accept the proposal. In return Briand informed Hughes that similar limitations on auxiliary ships would be impossible.[20] This led to further disputes over the size of cruisers, the outlawing submarines, ratios for aircraft carriers, and even discussions to limit airplanes and merchant marine shipping.

In the end, only limited auxiliary limitations were adopted: aircraft carriers were limited to a 27,000-ton standard displacement and also in a 5–5–3 ratio with a maximum total tonnage for this class not to exceed 135,000 tons (see Appendix 1). These limitations, when combined with the ten-year moratorium on building battleships inadvertently turned the aircraft carrier into a capital ship in its own right. Cruisers were limited to 10,000 tons per vessel with a maximum caliber of eight inches for the guns. All the participants agreed that unrestricted submarine warfare was illegal, but Japan and the United States argued strongly for retention of submarines, stressing their value as reconnaissance and screening assets for their fleets.[21] Senator Henry Cabot Lodge, a delegate at the Conference, outmaneuvered any opposition to the treaties in the Unites States Congress by pushing them through the Senate without formal hearings—to include the testimony of Navy officers.[22]

Historians tend to identify the failure to impose a comprehensive limit on auxiliary tonnages as the most significant shortcoming of the Washington Naval Treaty, since competition was moved to other categories of ships. The displacement of naval competition to other classes of ships focused navy officers on these classes and less on battleships. As discussed, this did not come easy to the Navy's officers. Denied traditional solutions—battleships and bases—the Navy looked at innovative new ship designs and concepts, particularly those for aircraft carriers, submarines, and cruisers. The failure to cap cruiser construction also worked in Japan's favor—with no limits on auxiliaries Japan was free to build as much as she could to further ameliorate the "inferior" position conferred by the ratio for capital ships. Most Navy officers in 1922 thought they could keep ahead of the Japanese in auxiliary construction (they were to be proven wrong). On the other hand, the fortification clause occupied center stage in the Navy's collective response to Congressional ratification of the Washington Naval Treaty (March 29, 1922). Solving the problem of a naval war in a theater devoid of forward bases dominated the thoughts and actions of naval officers (to include the Marines) for the next two decades.

The Fortification Clause: A Sea Power Problem

Navy leaders of this period used the concept of sea power to frame both strategic and operational problems. This concept rested on the premise that the "amount" of sea power a nation had depended on the size of the fleet, the size of its merchant marine (which reflected its maritime trade), and the number and location of naval bases and commercial ports. Using this concept in the aftermath of the signing of the Washington Treaty, these officers concluded that the fortification clause posed a serious challenge to their ability to apply sea power in defense of U.S. interests in the Far East—specifically, the defense of the Philippines and the "Open Door" trade policy in China.[23] The fortification clause contributed to innovation by forcing the Navy to consider how to apply sea power at very long ranges in the absence of preexisting bases. Planning and designing for this "worst case scenario" in turn led the Navy to innovate and develop differently than might otherwise have been the case if fortifications and base improvements had been allowed. The Navy at first sought traditional means to address the problem through the modernization of battleships. At the same time, the Navy considered and developed novel strategic and operational concepts to help it project power in the absence of bases.

A. T. Mahan's conception of sea power and what the Navy's leadership thought constituted sea power were not exactly the same. However, there was enough congruence between the writings of Mahan and the interwar Navy on the issue of sea power to constitute a consistent and agreed common understanding—what Dudley Knox termed "a concrete, comprehensive and coherent *conception* of modern war" (emphasis original).[24] Mahan's treatise *The Influence of Sea Power upon History* proposed that sea power was affected by "six principal conditions" that had to do with a nation's geography and national character. Throughout, Mahan emphasized the role played by having suitable harbors and maritime access, and the danger in not having a navy to protect them: "the question is eminently one in which the influence of the government should make itself felt, to build up for the *national navy* which, if not capable of reaching distant countries shall at least be able to keep clear the chief approaches to its own" (emphasis mine).[25]

In later years Mahan dealt in a more structured fashion with the fleet, merchant marine, and bases as elements of sea power. These elements were addressed in Mahan's lectures over the years at the Naval War College and published as *Naval Strategy* in 1909. Mahan saw sea power as essential to the foreign policy of the United States: "we can scarcely fail to see that upon the sea primarily must be found our power to secure our own borders and to sustain our external policy, of which at the present moment there are two principal elements; namely, the Monroe Doctrine and the Open Door."[26]

After the United States had acquired substantial overseas possessions Mahan addressed the issue of overseas bases in a new light vis-à-vis sea power. "If a nation wishes to exert political influence in . . . unsettled regions it must possess bases suitably situated; and the needs of commerce in peace times often dictate the necessity of such possessions which are acquired . . . when opportunity offers."[27] The members of the General Board of the Navy referred to these ideas and policies as the justification for a robust building program to build the treaty navy. The policy of the Open Door in China was cited specifically as being untenable in the absence of both western Pacific bases and a fleet with a 5–3 superiority over the Japanese fleet in all categories—not just carriers and battleships.[28]

Mahan addressed bases overseas in a somewhat Jominian fashion, as decisive points, or in his words "strategic positions."[29] Decisive battles in the modern naval age depended on the ability of a fleet to logistically support itself, particularly in remote areas with limited or inadequate basing. Mahan pointed to the lessons of the Battle of Tsushima to support these ideas.[30] It seemed self-evident to Mahan that these positions were a fundamental element of sea power just as they were a fundamental element in war. Their protection, acquisition, and use for offensive or defensive action, or the opposite set of actions against those of the enemy, were matters of great importance. Mahan devoted an entire chapter in *Naval Strategy* to the topic of defending and using "strategic positions" in a naval campaign, especially those that encompass overseas bases.

That Mahan's views on overseas bases were highly regarded by the Navy leadership is illustrated by an excerpt from a special supplement to the *United States Naval Institute Proceedings* in November 1921—the same month the Washington conference convened—written by President of the Institute Adm. Bradley A. Fiske arguing for the fortification of Guam and the Philippines: "In the chapter on 'Naval Bases' in my book, 'The Navy as a fighting machine,' page 322, is the paragraph 'Mahan states that the three main requirements of a naval base are position, resources and strength.' Now the book itself was published more than five years ago, and Mahan's declaration had been made some years before. The opinion expressed has never been challenged to my knowledge. It seems, in fact, to have been definitely accepted."[31]

The Navy's intended course of action prior to the Washington Naval Treaty, as articulated by Fiske, was to build a navy "second to none" and to fortify and expand its bases in the Far East. A fleet well-supported by bases demanded an entirely different set of design considerations from one that didn't. The fortification clause thus presented the Navy with both strategic and operational-level problems to solve in the western Pacific in the event of war with Japan.

In addition to sea power concepts, the Navy's pragmatic approach to the development of strategy also played a role in how innovation occurred during the interwar period. Navy strategic problem-solving was structured around framing problems and then testing their solutions. This method was "nurtured" at the Naval War College (NWC) by teaching naval officers how to write "estimates of the situation" and then using war gaming to test their assumptions and courses of action. The same method was used in drafting the naval portions of the colored war plans (Orange, Black, etc.) by the OpNav War Plans Division. Every plan began with an estimate of the situation. This method saved time, money, and probably lives. Strategic and tactical situations could first be addressed conceptually without risking men or ships and tested in an intellectual environment. Open discussion, criticism, and self-criticism were conducted after the completion of a "problem."[32] Adm. William S. Sims articulated this method in his welcome address in 1919 at the reopening of the NWC:

> The college aims to supply principles, not rules, and by training, develop the habit of applying these principles logically, correctly, and rapidly to each situation that may arise. This habit can be acquired only through considerable practice, hence the numerous problems in strategy and tactics. The process shows the necessity of:
>
> 1. A clear conception of the *mission* to be attained.
> 2. An accurate and logical *estimate of the situation*, which involves a systematic mustering of all information available and a discussion of its bearing upon the situation under consideration.
> 3. A *decision* that is the logical result of the *mission* and the *estimate*. (emphases original)[33]

This methodology was used to address the issue of the fortification clause in crafting an estimate of the situation for the 1924 Navy version of the Orange War Plan (WPL-9). The plan stated the relationship between sea power and mission: How would the U.S. Navy establish sea power in the presence of the Japanese Fleet in the Western Pacific? The collaboration between the planners at OpNav and the students and the faculty at NWC was extremely valuable in helping the Navy wrestle with the strategic challenge posed by the fortification clause.[34]

Having an interactive process that tested and retested planning assumptions for fighting a war in a remote theater without naval bases forced an entire generation of naval leaders to literally recalibrate their view of sea power. Innovations mean very little if they are not practical or logistically supportable. The dynamic strategic problem solving used by the Navy may not have directly caused innovations, but it surely tested them. When plans and designs were then forwarded for exercises by the fleet and consideration by the General Board, an additional layer of checks and balances was added to the inherent value of this system. A force structure was designed with strategy and operational capability in mind.[35]

Navy leaders, many of whom only learned the terms of the Washington Naval Treaty after it was signed, recognized the strategic problem posed by the fortification clause. The Navy's concern about the clause would fuel its efforts to innovate.[36] Capt. Dudley W. Knox, of the U.S. Navy, identified the impact of the clause in his book *The Eclipse of American Sea Power* in 1922 as soon as the final terms of the Washington Naval Treaty were made public. Knox was an extremely influential officer and still worked in the Navy (despite his retirement) as the head of the historical section of OpNav (later the Naval Historical Center).[37] He singled out the "sacrifice we have made respecting Western Pacific bases" as being of "greater importance" than the loss of tonnage strength embodied by the scrapping, tonnage limits, and the capital ship building "holiday." Knox also argued that without bases in the western Pacific ". . . we must build larger forces than hers [Japan] if we are to retain the ability to cross the Pacific without the aid of sufficient base facilities." Knox also argued that any attempts by the United States to increase the cruising radius or numbers of auxiliary ships required design trade-offs, and would be matched by unrestrained building programs by the Japanese who could economize given the proximity of their own bases.[38]

Navy planners took the fortification clause so seriously that they completely revised the Orange Plan for war with Japan. Knox had pointed out that the mandate islands (the Palau, Caroline and Marshall groups) lay across any American line of advance to the Philippines and Guam.[39] Since Japan controlled these islands and was closer to these islands than the United States, she could fortify them to a significant degree before the arrival of the U.S. Fleet. As the Orange planners were soon to posit, an approach that did not neutralize these islands would invite failure. In all likelihood, the United States would lose its weakly protected bases, as Knox predicted, and have to advance slowly through the Japanese-controlled waters, seizing bases as they went "step-by-step."[40]

Obsession with the fortification clause was not limited to Knox. Shortly after ratification of the Washington Treaties, retired Navy Adm. H. S. Knapp, a recipient of the Navy Cross in World War I and former member of the General Board, made the fortification clause the subject of his speech to the American Society of International Law at their convention in 1922: "Regarded from the viewpoint of position—comparative distances—Art. XIX is glaringly inequitable to the United States. . . . Art. XIX impairs for the United States the 5–3 ratio of floating strength with Japan in so far as the Western Pacific is concerned. The United States has yielded the possibility of naval equality in that region."[41]

This debate did not die down despite the Treaty being in force. Secretary Hughes felt compelled to respond to these criticisms when he addressed the annual meeting of the American Historical Association at the end of 1922:

> Failing to find unfairness in these provisions [tonnage limits] of the Treaty, there has been some criticism of the agreement to maintain the status quo with respect to fortifications and naval bases in the Pacific Ocean. [Hughes reads Article XIX verbatim]. . . . Was it not better that at a time of considerable tension, instead of threatening Japan by a proposal to fortify Guam, we should agree that for fifteen years we should rest content with the situation with which we had been satisfied for the past twenty-three years? And it should be remembered that in the same treaty Japan undertakes to maintain the status quo in the Kurile Islands, the Bonin Islands, Amani-Oshima, the Locchoo Islands, Formosa, and the Pescadores, and any other insular possessions she may hereafter acquire.[42]

Hughes' arguments did not put most Navy minds to rest. There were exceptions. For example, Adm. William V. Pratt, a new member of the General Board, was an outspoken supporter of the Treaty. Pratt's first direct comments were in an article he wrote in 1922 which addressed Knox's criticism of the fortification clause: "Much as the writer [Pratt] dislikes being in disagreement with some able writers on naval subjects, he cannot entirely follow their reasoning relative to Article XIX of the Treaty for a Limitation of Naval Armaments, nor accept as altogether sound their opinions of the status quo clause." Pratt did agree with Knox on one point, namely, ". . . to bend every energy to make our navy second to none."[43] How to build a navy "second to none" when neither Congress nor the president authorized or allocated money for its construction impelled Navy leaders to seek innovative solutions. They might not build, ship for ship, a Navy "second to none" for the Far East, but they might be able to find other means to redress the imbalance due to the fortification clause using new technologies, better tactics, and innovative operational ideas.

The General Board Implements the Treaty

As debates over the Washington Treaty raged the General Board convened and began to implement the provisions of the Treaty before it was even ratified. In this case the Board convened the first in a number of hearings with the objective of promulgating a formal Navy Policy "based on Treaty for limitation of Naval Armament" (See Appendix 2). The first meeting convened on March 3, 1922—the Treaty would not be ratified by Congress until the end of March (See Appendix 1). This meeting illustrates the range of participation, the influence of the treaty system, and the organizational dynamics of a General Board hearing.[44] The attendees at this meeting included the Executive Committee composed of Rear Admirals William Rodgers, Harry Huse, William V. Pratt, Major General John Lejeune, Captains Frank Schofield and Luke McNamee, and the Secretary of the Board (Commander M.K. Metcalf). Also present were representatives from the fleet and most of the bureaus.[45] This list shows how a hearing could have naval officers of varying experience, rank, and expertise at its disposal when addressing a particular topic. Missing was Rear Admiral Sims, President of the Naval War College. This was not unusual since Sims' normal duty station was Newport, RI. Sims often forwarded his opinions on topics of interest by letter. These were either read into the transcripts or attached as an appendix and then cited by the Board in its discussions.

As noted, the senior member of the Board on this date was Admiral William L. Rodgers. Rodgers had been present as a naval advisor to the U.S. delegation at the Washington Conference and used this occasion to speak at length about the profound changes occasioned by the Treaty:

> The Secretary of the Navy has directed the General Board to submit recommendations along the lines of the development of naval policy under the restrictions imposed upon us by the recent *Conference* [Washington], which reduces the probability of war for a number of years, and also reduces our building program and makes it necessary to maintain efficiency during a period of peace, which will in all probability last for some years. We can do nothing now, except under the terms of the *Conference*. We have a somewhat limited power of action under the terms of the *Treaty*. The Navy Department no longer can decide for itself on its course of action. Before even Congress can decide on a course of action it must come within the terms of the *Treaty* and for that reason it is desirable for the entire Navy to review its situation both internally [domestically] and with regard to the people, and externally with regard to other nations. These hearings afford a convenient method for us all to get together and acquaint ourselves with the various limitations which are imposed on us by the *Treaty*. In view of the appropriation bill and current legislation, it is desirable that we should clear our minds, all of us, as early as practicable, and for that reason I have asked the different Bureaus to send representatives here a little earlier than is desirable for giving them full time to study these sheets [these were the "blueprints" of the Naval Policy] which

we have sent out for discussion. We will listen to the Bureau of Aeronautics this morning. It is most desirable to have representatives of the other Bureaus present. . . . They are not only directly concerned, but we must tie up with the terms imposed upon us by the *Conference*. We have here the recommendations and suggestions from the Chief of the Bureau of Aeronautics [Admiral Moffett] and the Assistant Chief [Capt. Hank Mustin] is here personally. I will ask Captain Schofield to carry on with regard to the recommendations which at present concern capital ships and aircraft carriers and building programs. (emphasis mine)[46]

Admiral Rodgers' opening comments were unusual for their length, highlighting the seriousness of these hearings. The hearing then went on to discuss the first issue, the "Air Policy," that was to be included as a subcomponent of the Naval Policy (see Appendix 2).[47]

Admiral Rodgers' remarks show that he clearly recognized both the typical and exceptional features of this hearing. He took particular care to emphasize the traditional role of the bureaus in the process and how this meeting departed from the norm by bringing the bureaus' representatives into the deliberations earlier than usual. Once Rodgers was done with his introductory remarks he turned the facilitation of the hearing over to Captain Schofield. This was somewhat unusual in that a meeting of such gravity might normally be facilitated by the next senior admiral. Rodgers' choice of Schofield may in fact have been based on a preference for an officer of broad experience and potential. For example, Schofield was highly qualified and was promoted to admiral in February of 1924 at the age of 54, later becoming head of the War Plans Division of OpNav and later Commander-in-Chief of the U.S. Fleet (CINCUS).[48]

The impact of the Treaty is apparent—Rodgers emphasized it six times using the words "conference" and "treaty." Note, too, that Rodgers recognized that a paradigm shift had occurred and that all must "clear [their] minds" for deliberations from that point on. Rodgers observed, "The Navy Department no longer can decide for itself on its course of action." Rodgers made it clear that the Navy must answer to both domestic and international constituencies under the constraints of the Treaty. This passage also shows the intersection of the Treaty with budget considerations as the Board formulated overarching naval policy, including its constituent parts (such as air policy). Finally, the passage provides a view of the General Board as a sort of organizational "commons" where the bureaus, the fleet, and the senior leadership met to translate policy into concrete plans of action at the behest of the Secretary of the Navy.

Also noteworthy is the Board members' use of a preexisting document, for this hearing a draft "blueprint" for the overall naval policy.[49] The blueprint was augmented by written memoranda to the Board from various commands that could not be present (such as Sims at NWC). These "memoranda for the General Board" recommended paragraph by paragraph changes to the serial's wording or made general comments on the topic. The principal memorandum in use on this day was from Adm. William Moffett of BuAer and was duly attached at the end of the March 3, 1922 transcript.[50]

At this particular hearing, as the discussion proceeded, the Board's secretary made annotations or changes in wording to the proposed naval policy at the direction of Captain Schofield. For example:

> Captain Schofield: If there are no comments, I will assume approval. The next item is [from the Naval Air Policy]: "To delay building to tonnage allowance until new aircraft carriers shall have been tested in service. Then to build aircraft carriers to total tonnage permitted by Treaty." That has all been discussed this morning as a result of the remarks of the Chief of Aeronautics.
>
> Commander Leary [Bureau of Ordnance]: There is a comment from Admiral McVay on that point.
>
> Captain Schofield: [reading from Admiral McVay's memorandum] "It is recommended that the words 'to delay building to tonnage allowance until new aircraft carriers shall have been tested in service. Then' be omitted, as these words have a tendency to throw doubt upon airplane carriers as contemplated at present, whereas testimony before the House Naval Committee was to the effect that improvement might be expected in accessories rather than in the vessels themselves."[51]

Schofield was running the hearing with little interference from Rodgers and the bureaus' representatives had no qualms about speaking up. The informal way in which Commander Leary addressed Captain Schofield highlights that the normal rules of protocol with respect to rank were relaxed a bit in the hearings in order to promote a more open atmosphere conducive to discussion. The occasion for this discussion, testing carriers operationally in the fleet, emphasizes the Board's (and the Navy's) willingness to experiment. Admiral McVay was making the point that the purpose of the testing was not to "prove" the value of aircraft carriers to the Navy, but rather to improve its efficiency for the fleet.[52]

The closing of the March 3, 1922 hearing emphasizes the next steps that were taken in the hearing process on the naval policy as well as the way in which the

Board typically terminated a hearing. The closing summarized what had been addressed at that day's hearing as well as provided the participants direction on their role during the remaining hearings on naval policy:

> Admiral Rodgers: . . . I would like to ask who is the senior man here from Operations [CNO]? Would you be ready to come here on Monday and answer questions or to give your views on this?
>
> Captain Overstreet: On the whole group?
>
> Captain Schofield: Except the air policy. I don't think we want to go over this again.
>
> Admiral Rodgers: No, the Air Policy is finished.
>
> Captain Overstreet: Yes, sir. You just mean to continue the discussion of the other features.
>
> Admiral Rodgers: Ordnance [BuOrd] comes in under the terms of the Treaty. [because of this] It [BuOrd] has nothing very new or original [to recommend for changes to the naval policy], as it seems to me Ordnance will come in as criticizing and suggesting [to] all the others. Will you be ready on Monday?
>
> Commander Leary [Bureau of Ordnance]: Yes, sir.
>
> Admiral Rodgers: At this present stage, I don't think it is advisable to ask Ordnance to have a hearing on anything special.
>
> Captain Schofield: We might allot on Monday a certain portion of this schedule and ask the bureaus and offices to be present to discuss it. For instance, say the General Naval Policy, and discuss that.
>
> Admiral Rodgers: I think that is a good idea. Take up the General Naval Policy and if we get through that quickly we can go on with the rest as fast as we can. We are much obliged to you all for your help.[53]

By the meeting's end the dialogue had switched from the topic at hand, the naval air policy, to the next set of hearings. Rodgers and Schofield identified not only the next topic, the General Navy Policy, but the general role that all the bureaus were to play for the remainder of the hearings on this topic. In particular, they focused on BuOrd assuming the role of a critical observer with regard to the suggestions of the other bureaus and offices. By doing this, Rodgers had elevated BuOrd's status to the level of a temporary member of the Board for the hearings on the new naval policy. By the end of 1923 the new *Naval Policy of 1922* (the first naval policy document of its kind) had been published and disseminated throughout the Navy as the "blueprint" the Navy intended to adhere to as it implemented the terms of the Treaty.

Despite the passage of time and their rather energetic implementation of the terms of the Treaty, Navy officers bemoaned the fortification clause in a variety of forums. Pratt's friend and fellow member of the General Board, Capt. Frank Schofield articulated the fortification clause's impact on sea power to an Army War College audience in September of 1923: "*Sea Power* is not made of ships, or of ships and men, but of ships and men *and bases far and wide*. Ships without outlying bases are almost helpless—will be helpless unless they conquer bases and yet the Treaty took from us every possibility of an outlying base in the Pacific except one [Hawaii]; we gave our new capital ships and our right to build bases for a better international feeling—but no one gave us anything. *Manifestly the provisions of the Treaty presented a naval problem of the first magnitude that demanded immediate solution.* A new policy had to be formulated which would make the best possible use of the new conditions." (emphasis mine)[54]

How the Navy framed strategic problems during the interwar period should be viewed with the understanding of the impact of the fortification clause on sea power in mind. As mentioned, Schofield later went on to become the head of the OpNav War Plans Division responsible for drafting the Orange War Plan. He also served as Commander-in-Chief of the U.S. Fleet (CINCUS). As such, he wielded considerable influence throughout the life of the treaty navy. Schofield identified a clear problem and his last sentence provides an explanation for the path the Navy chose to address this problem—a policy that directed the ship designs that would also ameliorate the fortification clause. The fortification clause subsequently became a central consideration in the practice problems and war games at the Naval War College. It also served as the dominant factor in the planning for an Orange War by the War Plans Division.[55]

By 1930, amidst the angst of the London Naval Conference, the sacrifices made as a result of the fortification clause caused the reformer Admiral Bradley A. Fiske to wire the General Board that "In 1921 General Board was planning to prevent war with Japan by *adequate defense of Philippines* but the conference of 1922 reversed that plan against protest of [the] Board and that this *has been the bottom cause of all subsequent troubles*." (emphasis mine) [56] The Navy and the General

Board continued to see the clause as the source of its problems right up until the Japanese left the treaty system at the end of 1936. Among the Board's first actions after the Japanese left was to draft a serial for the Secretary of the Navy pointing out that the limits of the fortification clause no longer applied. They warned that "... independent negotiation of Article XIX, modified or unmodified, would be detrimental to the interests of the United States." They preferred that the article lapse along with the rest of the now-compromised Washington Naval Treaty.[57]

The U.S. fleet in the Far East would also have to meet the new requirement of operating with air cover, a problem of considerable importance in the absence of land-based aviation. Also, the Navy was forced to consider how its fleet might support the seizure and establishment of new bases, or the relief or recapture of the few existing ones. These were problems that it had only just begun to seriously address when the first treaties were signed.[58] Navy leaders understood that the fortification clause was the critical element constraining their strategy in a war with Japan. Thus it contributed most in influencing innovation, principally by changing Navy leaders' attitudes. They were now more likely to champion ship designs for a long-range, self-sufficient fleet.[59]

* * *

Of the twenty-three articles in the Washington Naval Treaty, Article XIX addressing the status quo of fortifications in the Pacific had the most impact on innovation and ship design in the U.S. Navy during the interwar period. Its impact was greater than that of tonnages, ratios, or "naval holidays." This is because the fortification clause affected the probable maritime theater of war more than any other factor across the levels of war. While tonnage factors influenced the sheer numbers of capital ships and carriers that could be built, the ratios tended to minimize operational and strategic design factors—they were already fixed. The building holiday, too, influenced the types of ships which could be constructed, but it did not set the context within which the "frozen" capital ship tonnage and its supporting auxiliary tonnage would be employed. This context came from the fortification clause. Navy leaders during the period built toward parity with Great Britain as the goal but with Japan as the target of their strategic focus and most likely adversary in a future war.[60] Because of this, ships were designed for operations in the far reaches of the Pacific. The fortification clause also influenced the Navy's application of sea power more than any other article because it forced the Navy leadership to consider how to solve its lack of basing with a treaty fleet. It is hard to imagine the passage of the "Two Ocean Navy Act" in 1940 in the absence of this mindset. Similarly, it is equally hard to imagine the early success of the treaty fleet in World War II if the leadership truly was obsessed with battleships instead of the sustained sort of power projection needed in the absence of forward bases.

Chapter 4

The General Board and the Treaty System

The Navy attempted to solve the challenge posed by the fortification clause in a number of ways, all of which contributed to innovation. The Navy had a standard method for problem-solving that it taught and practiced at the Naval War College. Additionally, the OpNav staff used the fortification clause as an integral consideration in its planning process for Orange courses of action by its War Plans Division. These courses of action were then forwarded to the fleet, the Army-Navy Joint Board, and the Navy War College for testing and refinement. Officially called fleet problems, these were investigated during annual fleet exercises that were usually held in the summer to address new tactics, techniques, and platforms. For example, Fleet Problem IX (1929) addressed the tactical and operational use of the new aircraft carriers *Lexington* and *Saratoga* within the strategic context of a defense of the Panama Canal. The Navy addressed the fortification clause's constraints through the General Board. The Board applied solutions by setting the criteria for the design of ships. The Board also exercised its influence by recommending budget priorities for the Secretary of the Navy.

As discussed in the previous chapter, the General Board served as the central authority of the U.S. government for implementing the provisions of the Washington Naval Treaty. The Board's members attempted to leverage the system by liberally interpreting the various clauses of the Washington and subsequent treaties. They attempted to wring every last advantage allowable or implied by the treaties—often without consulting the State Department. For example, Board members believed improving propulsion systems on battleships did not constitute "reconstruction" as defined in the Washington Naval Treaty (Article XX, Part 3). The chairmen of the executive committee of the Board also attempted to achieve advantages within the treaty system by altering or adding language to proposals that were favorable to the strategic situation of the United States. This would be the case with an innovative new ship design that came to be known as the flying deck cruiser in 1930. The General Board was sometimes frustrated in its efforts to leverage the treaty system during the period, tending to lose rather than gain ground. This, in turn, contributed to innovation by forcing Navy leaders, especially the members of the General Board, to address problems differently than they had in the past. Building more battleships and enhancing or expanding overseas bases were no longer options. Even the money to maintain the existing facilities was cut back during the fiscal crises of 1931 and 1932.[1]

* * *

Prior to March 1922, the General Board had prepared its building recommendations in accordance with Navy regulations. The result was an annual naval construction program summarized as a General Board serial. For example, Serial No. 1055 written in 1921 was for naval construction in 1923. Prior to the Washington Conference, these building programs reflected the policy of naval parity with Great Britain first articulated in 1915 as a "Navy second to none." The Board's members thought that the Washington Naval Treaty had codified this policy once and for all. Accordingly, they took advantage of the opportunity afforded by the Treaty to support their building recommendations by tying the "second to none" axiom to the Washington Naval Treaty in the 1922 *U.S. Naval Policy*. This policy was to be distributed to the entire Navy to educate it with the terms of the Treaty and its impact. The *Naval Policy* was to be reviewed on an annual basis, but the Board hoped that no substantive changes would be necessary in the short run given that the Treaty was fifteen years in duration.[2]

The Board's first step in crafting this policy had been to canvass Navy organizations for their input. The fleet, the bureaus, and the War College were among the most important organizations queried for their inputs on the "tentative draft U.S. Navy Policy." Since the CNO was a full-time member of the General Board, he provided OpNav's inputs directly. As previously discussed, Admiral Moffett, the chief of BuAer, responded to the Board on March 2, 1922. Moffett's response indicates the broad scope of inputs allowed by the hearing process of the Board. For example, the BuAer chief did not limit his recommendations and opinions to air matters but commented on all classes of ships. The first subparagraph of his memorandum urged highest priority for modernizing battleships. The other Bureaus responded similarly.[3]

The Board's first paragraphs in the finalized serial on the new policy were philosophical in tone. Despite applauding the sentiments that led to the success of the Washington Conference, the Board recommended building to the absolute limits allowed by the treaty. They further recommended that this construction program be retained until "serious evidence" of a "spirit" of mutual "confidence and frankness" presented itself. This language was not what the cost-cutting Harding administration wanted to hear. It highlights the fundamental difference of opinion between the interwar General Board's leadership and their civilian leadership regarding naval arms limitation. The administration's budget strategy assumed that good faith generated by the treaty system would lead to further arms limitation across the military spectrum. The best way to maintain this momentum was to have more conferences, which in turn would result in reduced naval construction. In order to show good faith, current naval construction should be delayed or even put on hold.

The General Board took the opposite position. For these officers, good faith must be demonstrated by the other parties to the treaties before considering any relaxation of a naval construction program that aimed at achieving the 5–5–3 ratio and a Navy "second to none." The General Board wanted to build every ton permitted and build it quickly. The Board's skepticism toward the Treaty is exemplified by the last sentence of its serial on *U.S. Naval Policy* of May 1922: "The General Board recommends therefore, that steps be taken to keep the uncompleted 1916 capital ships in a state of complete preservation and that plans be in readiness immediately to resume work on this program should it become evident that the treaty will not be ratified within a reasonable period (one year from date of signature) by the signatory powers, and that meanwhile no old capital ships which are still useful be scrapped."[4]

As further justification, the authors of this serial cited the Navy's explicit and overarching role in protecting the coasts of the United States, overseas interests, and seaborne commerce. The authors then addressed the particulars of implementation of the Treaty with respect to naval construction. Most importantly, this serial recommended that the 5–5–3 ratio "be made the present basis of building effort *in all classes of ships.*" (emphasis mine)[5] This was consistent with the Board's goal of rectifying the failure to limit auxiliary tonnage by unilaterally observing auxiliary limits and attaining these ratios by the next naval arms limitation conference. By doing this, the Board would have a benchmark for building auxiliary classes of ships should Japan, for example, start building large numbers of cruisers. The goal, reflected in the *U.S. Naval Policy* serial, was to achieve these ratios for modern cruisers by laying down eight of this class by 1924. This move was justified as "replacement" construction to make it more palatable to the public. However the foremost priority budget item in the 1922 recommendations for the 1924 fiscal construction plan was the modernization of the existing battleship fleet.[6]

Aircraft carriers were a different story. The Washington Naval Treaty allowed both Japan and the United States each to convert two capital ships already under construction into large aircraft carriers. This gave the United States and Japan approximate parity in aircraft carrier tonnage, not including the experimental aircraft carriers *Langley* (U.S.) and *Hosho* (Japan). This meant that the United States was already *behind* in its 5–3 ratio allowance for carriers. To rectify this problem, the *U.S. Naval Policy* serial recommended that construction begin at once on a 27,000-ton aircraft carrier. One can see that, as predicted by Dudley Knox, naval building competition had already shifted from battleships to cruisers and carriers. The United States also used the Washington Naval Treaty as a pretext to build the most modern ships permissible.[7]

The modernization of existing capital ships along with an increased focus on non-capital ship design represents one path that innovation followed in response to the treaty system. Modernizations in the U.S. Navy meant increasing the efficiency of existing ships, and in this case the battleships. In this sense, the battleship modernization of the interwar period can be characterized as evolutionary, not revolutionary, innovation. Too, the Treaty had forced the Navy to more seriously consider the design of those ships categorized as "auxiliaries." With less of the executive intellectual and organizational effort focused on battleship building and design considerations, the Board and the bureaus now gave more of their attention to other classes of ships. Additionally, the switch in focus tended to enhance the Board's, and the rest of the Navy's, tendency to investigate and develop different doctrinal ideas and ship designs.

There is a saying that "work expands to fill the time allotted to it."[8] This was partially the case with the Navy, and especially the General Board, during much of the interwar period and was an unintended result of the capital ship holiday. Discussion that might have otherwise been devoted to new designs for battleships and strategies for implementing their replacement was switched to new areas. Once the Board's recommendations for modernizing the battleship fleet were approved, these ships occupied relatively little of the members' intellectual attention. These officers were still required to defend the modernization program and ensure that it was funded, but they had ample time on their hands to address the pressing issue of staying ahead of the Japanese in cruisers, aircraft, carriers, and submarines. The Board could also focus on creative logistics workarounds to the fortification clause. Examples of these included ideas like submersible fuel tenders, mobile bases, and floating dry docks.

Within the 1922 *U.S. Naval Policy* was a section entitled "Building and Maintenance Policy" that reflected the impact on naval construction and innovation:

- To make the capital ship *ratios* the basis of building effort in all classes of fighting ships.
- To make superiority of armament in their class an end in view in the design of all fighting ships.
- To provide for *great radius of action* in all classes of fighting ships. (emphasis mine)[9]

It is clear that the Navy wanted to ensure that it had a written basis for continued naval construction to keep up with Great Britain and ahead of Japan. Secondly, this policy emphasized innovation in two areas, "superiority of armament" and cruising radius. The Navy regarded superiority of armament (meaning as large a caliber of gun and as much armor as possible) a priority requirement in the design of its ships. The final general design precept of the building policy emphasized the new importance the Navy placed on building ships with efficient power plants and large fuel reserves given the lack of shore-based logistics support in the Pacific. This especially showed the effect of the fortification clause. The efforts to maximize protection (armor) and striking power (guns and eventually aircraft) conflicted with the requirement for maximum storage space for provisions and fuel for long-range, sustained cruising. The tension between these design characteristics immediately led the Navy to advocate building to the limits of individual ship sizes—10,000 tons for cruisers, 27,000 tons for aircraft carriers, and large "fleet" submarines. These details were spelled out immediately following the general precepts of the building policy (see Appendix 2).[10]

The Navy's ambitious plan to build a treaty fleet immediately ran afoul of presidential and congressional economizing. The General Board's membership believed that an absolute 5–3 ratio in all classes of ships must be built to. Once their policy was executed—in the form of a healthy construction program tied to the naval construction of the other powers—then the United States would be in a position of strength, as it had been prior to the Washington Conference. The diplomats then might effect further limitations at another conference. The Harding and Coolidge administrations held that the United States must set a good example by restraining naval construction. The Washington Naval Treaty had created the conditions that caused the administration to scale back the language of the 1923 *U.S. Naval Policy*.

The first harbinger of the administration's curtailment of naval construction occurred when Navy Sec. Edwin Denby convened the General Board to discuss the language for the next year's (1923) *U.S. Naval Policy* serial. In particular, language that tied building to the specific tonnages of other foreign powers was deleted. Further, the Secretary eliminated the words "to maintain the present extent and efficiency of all naval facilities now existing in Guam and the Philippines" as provocative. On December 4, 1922 the Secretary "orally" informed the Board that they were to strike the following clause from the "Allocation Policy" of the *U.S. Naval Policy*: "That the principal naval strategic center of interest in the world today is the Pacific." In this manner Japan was not to be provoked. These changes had the effect of undermining policy justifications used by the Navy for its construction programs.[11]

The Harding administration attempted to establish the basis for another arms limitation conference in Washington in 1924 in order to build on the perceived success—and rectify the shortcomings—of the first Washington Conference. Subsequent administrations of the 1920s held true to a policy that attempted to use the treaty system to enact further naval limitation. Presidents Harding, Coolidge, and Hoover reasoned that moderation in U.S. naval construction would spur the other major powers to agree to another conference. Such a conference, if successful, could make unnecessary the ambitious cruiser and carrier programs being advocated by the General Board.[12] President Harding's death in the summer of 1923 and the subsequent Teapot Dome scandal temporarily pushed naval limitation to the back burner.[13] But in 1924 President Coolidge focused the General Board on the issue of another limitation conference. He directed his acting Secretary of the Navy, Theodore Roosevelt Jr., to query the General Board about further naval limitation "as soon as conditions in Europe justify such a move."[14] Roosevelt outlined four areas that he wanted the Board to consider:

1. naval ratios, "if the same formula [should] be used,"
2. "methods" for limiting "military or naval aircraft,"
3. new "angles under which the submarine could be treated," and
4. merchant marine restrictions.[15]

The Board could drag its feet when it wanted to and it was happy to leave the Washington Naval Treaty unmodified, especially since the new initiatives implied further cuts and restrictions. The only serial on arms limitation between August 1924 and June 1925 was No. 1239. General Board records for that study show no action until June 3, 1925.[16]

Almost a year after Roosevelt's query, the president again revived the limitation process, directing the new Secretary of the Navy, Curtis Wilbur, to ask the Board again for its recommendations on the topic of naval limitation. The Board was now chaired by Adm. Hilary P. Jones, whom historian William Braisted has called "the Navy's recognized authority in arms limitation."[17] The Board's first priority was to recommend against any extension of the ten-year capital ship building "holiday." Allowing the "holiday" to end in accordance with the Treaty would enable the Navy to start replacing its oldest battleships with newer, more modern vessels. The Board also recommended retaining the ratio method "for all types not now limited and limitation of total tonnages in each." Admiral Jones took advantage of the occasion to emphasize once again the role that the Board's interpretation of sea power played in arms limitation: "As sea power comprehends *combatant ships* plus *merchant marine* plus *bases*, it follows that, in comparison with British sea power, for instance, if the United States and Great Britain arrive at an equality of combatant ships, Great Britain, by reason of superior strength in merchant marine and bases remains the dominant sea power." (emphasis mine)[18]

Secretary Wilbur was less than pleased with this advice and brusquely annotated the serial with the instruction to "Please reconsider this recommendation." Interestingly, Japan was not mentioned by name in the serial. The General Board had perhaps learned a lesson from previous experience when its focus on "Pacific strategy" had earned it a sharp rebuke and was now using Great Britain as the reason for taking a hard line in arms limitation. If so, this tack did not seem to work either.[19]

General Board serials reflected a deepening pessimism each year after 1922 about the prospects for building a full-strength treaty navy. In a December 1926 serial, the Board again emphasized the need to build to treaty limits because it was "the only fleet that will provide adequate national defense." However its assessment of the status of the existing fleet was blunt, "We have not such a fleet at the present time."[20] By the next year, as building continued to be delayed by both parsimony and hopes for further arms reductions, the Board's tone became more urgent: "... we have fallen behind and continue to fall behind the treaty ratio of 5:5:3 in certain classes of auxiliary combatant craft [cruisers, carriers, and submarines]."[21] Faced with no additional authorizations for new naval construction the Board (in 1927) was forced to trim its fiscal year 1929 building program to what it viewed as absolutely essential. The Navy's priorities incorporated elements of modernization within a program to build up to treaty limits. These priorities addressed U.S. strategic defense requirements through technological and operational innovation. The following list illustrates how the Board identified and prioritized the most important programs:

a. Modernize five oil burning battleships in accordance with previous recommendations of the General Board.
b. Lay down eight ten thousand-ton modern cruisers.
c. Lay down three fleet submarines.
d. Lay down one aircraft carrier not to exceed 23,000 tons.
e. Construct aircraft in accordance with five-year building program recommended in [the Bureau of Aeronautics memorandum].
f. Lay down one floating dry dock.[22]

These recommendations reflect a commitment to an innovative means of pursuing an enduring program that also accounted for the fortification clause and fiscal realities. Only the most fuel-efficient battleships were chosen for modernization in the following year. This way the longest "radius of action" ships received the first gun turret modifications and other combat efficiency upgrades needed for battle in the western Pacific. Likewise, large cruisers and submarines were given priority in order to support the battle fleet and provide long-range reconnaissance.

Floating dry docks for deployment to either the central Pacific or the Philippines reflected the work of the Orange planners and were first identified in the 1923 Board serial on naval building policy. The presence of this program was unusual. The 1924 Orange Plan included a plan for a "mobile base" as a secret annex. This plan outlined an ambitious program to build large floating dry docks which could be deployed to undeveloped, but promising, locations in the Marshall, Caroline, or even Philippine Islands in order to ameliorate the effects of the fortification clause.[23] Up to that time the General Board had not included floating dry docks among the warship programs. The floating dry dock's retention in the pared-down priority list above gave it an equivalency with the warship programs that was remarkable for that day and age.[24]

The emphasis on aviation also reflects the innovative mindset of the members of the General Board. However, the Board, desperate to build an aircraft carrier, scaled back the projected carrier's size to 23,000 tons (from 27,000) in the hopes that the smaller overall cost would meet the aims of Congress and the administration "to keep down appropriations . . . avoiding excessive peaks in expenditure in ensuing years." By November 1927, the Board had held hearings and decided that a small 13,800-ton carrier was better than nothing.[25] Carrier construction also illustrates the effect of the interaction of the budget process with the hopes generated by the treaty system for even more naval limitation.

It has been said that "the President proposes and Congress disposes." Actually the process was (and is) much lengthier and more complex. Once Congress authorizes naval construction the president then prepares a budget that may or may not include funding requests for whatever naval building has been authorized. Also, the president can presume spending authority in his budget even though Congress has not authorized it. Congress, however, can modify or eliminate these proposals in its spending bills, which of course the president can veto. Often in the interwar period ships were authorized but not built until much later. If a new arms limitation treaty reduced tonnages and sizes the ships might not be built at all. For example, the language in House Resolution 8687, which authorized naval construction in 1924, clarifies the relationship between construction and the treaty system: "That in the event of an international conference for the limitation of naval armaments the President is hereby empowered, in his discretion, to suspend in whole or in part any or all alterations or construction authorized in this Act."[26] The president's budget submissions could also be scaled back during the allocation process by Congress, whose membership also included arms reduction advocates. Legislators often saw no sense in building ships that might be scrapped under the terms of a new treaty as they had been after the Washington Conference.

The first carrier built after the conversion of the battle cruisers *Lexington* and *Saratoga* reflects precisely this process. It also reflects the attempts by the Navy to try and come up with a successful budget strategy focusing on the word "replacement." Adm. William Pratt coined it a "best seller."[27] The carrier size recommendations show how the budget process—influenced by the administration's plans for more arms limitation—caused the General Board to compromise over time in order to get at least one carrier of reduced size built.[28] These factors contributed to the extremely small size of the first U.S. aircraft carrier built from the keel up for that purpose—the USS *Ranger* (13,800 tons). Even so *Ranger* was not commissioned until 1934—twelve years after the first proposal to build it.

Admiral Jones was dispatched in 1926 to Geneva as the senior observer of the United States to the League of Nations Preparatory Commission for Disarmament despite the clear differences over naval construction between the Coolidge administration and the General Board. The League had established this commission to discuss the basis for a comprehensive arms reduction conference tentatively scheduled in 1927 at Geneva. Jones took with him the same agenda Secretary Wilbur had asked him to "reconsider" the previous year. In part this was due to the continued reticence of the Secretary to provide either formal approval of the original serial (1239) or new written guidance. Prior to leaving for Geneva, Jones had approved a response to a query from Admiral Pratt on this very issue: "No one has been able to get any information regarding the reasons why the Secretary declines to sign such a precept, nor have any specific modifications to it been suggested."[29]

Jones ably defended U.S. interests at the Preparatory Commission meetings. Jones' main purpose at the 1926 Geneva meetings was to maintain the primacy of the Washington formula in the face of French proposals to limit navies by using a "global tonnage approach." This method simply gave nations a maximum tonnage figure and they could build as they saw fit within it—a variation on this theme was to add merchant tonnages to that of the warships. Jones found willing allies in the Japanese and the British naval delegations for maintaining the Washington formula as the correct basis for future conferences—tonnages and ratios by class. Jones judged seeking a universal consensus from these meetings to be a "hopeless task." However, Jones and the administration held out hope for a successful effort in 1927 outside of the aegis of the League. The State Department called for a three-power conference to meet the next year at Geneva "independent of the League" in part due to Jones' recommendations.[30] Forewarned by the U.S. Naval attaché in London about British proposals for the conference, Jones emphasized the opposition of his country to any initiatives to abolish the battleship or the submarine.[31] The retention of the battleship has been viewed as evidence of the interwar Navy's obsession with this vessel. If one looks at this issue from the perspective of the General Board at the time, one finds instead that the battleship was viewed as the one element of the treaty navy that the United States had managed to maintain at close to parity in the wake of the Washington Naval Treaty.

As for the submarine, it was viewed as essential to strategy in the Pacific based on the courses of action in War Plan Orange. Captain Schofield, future chief of OpNav's War Plans Division, recommended using submarines as the "eyes" for a fleet denied sanctuaries in the western Pacific.[32] This was one of the reasons the Navy spent so much time and effort in designing its large "cruiser" submarines with 11,000 nm endurance and a planned patrol length of up to 75 days.[33] Jones had been a party to the extensive Board hearings on the topic of fleet submarines in 1924–1925 and regarded the submarine as an essential reconnaissance and screening asset for the fleet. This principle was now a component of the *U.S. Naval Policy* and reflected in tactical practice. During the hearings, Jones and the Board had focused on how to get as much "radius of action" as possible from these vessels, including holding hearings on a "submersible fuel tender." The Board's hearings on the submersible fuel tender highlight the intersection of new technology with innovative operational approaches in response to the geography of the Pacific and the constraints of the fortification clause. It also highlights the use of the General Board hearings as the preferred forum for the initial discussion of new concepts.[34] This idea foreshadowed the actual use of "milk cow" logistics submarines by the Germans for just this purpose during the later stages of the Battle of the Atlantic in World War II.

The submersible fuel tender also provides an interesting example of the influence of the budget process on innovation during the 1920s. This hybrid vessel was never built, even as a test bed. This was principally due to the limited funds, which were recommended and programmed for the most urgent priorities—battleship modernization, cruisers, and aircraft carriers. Also, the submersible fuel tender example highlights a shortcoming in the Navy's methods: A promising concept had been proposed—it was clear from the transcripts of the hearings on the submersible fuel tender that the General Board's members thought as much.[35] However, without a prototype vessel, the concept could not be tested in the fleet and the concept might be shelved or even forgotten. In the case of the submersible fuel tender this is actually what happened. By the time the issue was raised again, the General Board had decided to address the operational radius of its submarine fleet by simply building larger, more fuel efficient, and habitable submarines. However, money was slow in coming to construct these vessels (they also had technical difficulties) whereas the construction of even one submersible fuel tender would have immediately increased the operational range of the submarines already in the fleet as well as provided valuable design data through actual experimentation. In this case the budget process had adversely affected innovation instead of encouraging it as some historians claim.[36] Admiral Jones' involvement in hearings for the design of submarines on the one hand and treaty negotiation on the other is also a useful example of the intersection of innovation and the treaty system.

Admiral Jones was detached from his General Board duties again to visit his British Admiralty counterparts in November 1926. Jones and the British ended up talking past each other on the issue of cruiser tonnages and ratios. This was due to a British reluctance to reveal their overall cruiser requirements, which were substantially higher than the United States was willing to accept. The cruiser issue was critical. The British wanted to build large numbers of smaller cruisers (7,500 tons) to secure their imperial sea lines of communication. They wanted a cruiser ceiling that was double (600,000 tons) what the General Board had recommended. The General Board knew they would never get Congress or the president to approve a program to build up to 600,000 tons—in fact the General Board had calculated that 300,000 tons was probably the maximum they could get authorized for cruisers built and building. Jones continued his shuttle diplomacy, meeting with his British counterparts again in March 1927 to ascertain if they were more flexible on the cruiser issue than they had been the previous November. The British led him to believe they were.[37]

The American position was exacerbated by the fortification clause. Because of this clause the Americans were building cruisers to the treaty limit of ten thousand tons to give their cruisers as large an operational radius of action as possible in the absence of bases. The required range, as with the submarines, was a direct result of War Plan Orange requirements for the cruisers of the scouting fleet. The best way to get the most efficiency and range was build the largest cruisers allowed by treaty and with the most efficient diesel propulsion technologically possible. The Navy ended up making alliances with the U.S. Railroad industry in order to have them convert to diesel fuel for all their trains. In doing this they hoped to get the commercial sector to develop the best smaller diesel engines without the expense of having to do it themselves. The result of these machinations were the first true "treaty" heavy cruisers (CA) of the Pensacola class. They had over seven thousand nm of endurance steaming at 15 knots. However, these ships would have to be based at Pearl Harbor in order to execute any aggressive application of War Plan Orange (the "through ticket").[38]

The British, on the other hand, had less need of these large cruisers given their mature and extensive system of world-wide bases. The Americans could not afford to build to match British tonnages. The United States was also constrained by the size of the cruisers it felt it must build—the largest cruisers possible under the terms of the Treaty. Big, modern, well-armed cruisers were less dependent on bases.[39] The difference between the United States and British positions was profound—the gap separating the two positions was 300,000 tons. Jones and his civilian counterpart, Ambassador Hugh S. Gibson, proceeded to the Geneva Naval Conference in June 1927 with no idea that the difference was so great.[40]

The U.S. position going into the Geneva Naval Conference reflected both hope and regret. The hope was that a treaty agreement would restrict the other powers' cruiser construction programs. At the same time, Jones and the General Board regretted the negotiating position their country's anemic naval construction program had placed them in vis-à-vis Japan and Great Britain. At the time of the Geneva Conference the United States had begun construction on only two of the eight cruisers first authorized in 1924.[41] Great Britain on the other hand had laid down thirteen new cruisers and Japan eight. Clearly U.S. construction was lagging in this category, as the General Board had repeatedly emphasized in its serials. The U.S. plan was to propose a cruiser limitation of 300,000 tons for the United States and Great Britain and restrict the Japanese to 60 percent or less of this total (180,000 to 150,000 tons). However, the context for the Geneva Conference was completely different from that at Washington in 1921–1922. At the first conference, the United States had negotiated from a position of strength—capital ships built and building. However, at Geneva, it was in a position of relative inferiority on the numbers of cruisers. It was thus less likely that either Great Britain or Japan would agree to any substantial limits without significant concessions from the U.S. delegation.[42]

The Coolidge administration's desire to end, or at least slow down, the cruiser race ended once the British and the Japanese responded to the initial U.S. proposal. The British unveiled their proposal for a 600,000-ton ceiling on cruiser tonnage while the Japanese intimated that they regarded a 70 percent ratio as the minimum basis for discussions. Additionally, the U.S. delegation was thunderstruck by the revelation that the British had been measuring their capital ship displacement in legend tons instead of standard tons, which meant the United States was some 100,000 tons behind the British in capital ship tonnage. The U.S. delegation, Jones especially, accused the British of bad faith in the talks leading up to the conference.[43] The Navy could not get the new construction it wanted from its government but in turn the United States had not built enough to be able to negotiate in a meaningful way with the other powers. This sad chain of events pushed the Navy toward even more efficient designs and innovative operational concepts.

Some historians have criticized Jones' and the General Board's lack of flexibility, but the United States was willing to increase the cruiser tonnage limit for the British to 400,000. The Japanese delegation under the able leadership of Adm. Saito Makoto toned down its rhetoric for a 70 percent cruiser ratio vis-à-vis the United States in order to avoid a charge that they had scuttled the conference. The U.S. and British delegations remained at loggerheads over the maximum tonnage limit.[44] Given the U.S. track record in failing to build to limits, Jones and his delegation could not afford to go beyond the 400,000-ton compromise offered to Great Britain. Any increase in the overall tonnage would result in a proportionally higher tonnage that the Japanese would certainly build to. In the upshot, the Japanese position was overshadowed by the Anglo-American impasse over the tonnage ceiling. The conference became acrimonious in the extreme with the head British delegate Lord Cecil resorting to verbal abuse of Admiral Jones. Ambassador Gibson threatened to walk out with the entire delegation. The conference broke down not long after.[45] It is ironic that the two powers who would become the firmest of Allies in World War II regarded each other with such hostility.

The breakdown of the naval arms limitation process at Geneva prompted the Coolidge administration to take action. Taking advantage of the impression of bad faith projected by the British, and perhaps lingering Anglophobia, Coolidge proposed, and the House passed, a second "cruiser bill" shortly after the end of the conference. This bill was rejected in the Senate due to fierce lobbying by powerful peace groups. However, a modified "cruiser bill" was submitted in 1928. This bill authorized the construction of fifteen additional cruisers and a small aircraft carrier. Even so, the Senate delayed approving the bill until after the 1928 fall elections.[46]

Absent from this new bill were any funds for the construction of a floating dry dock. However, the General Board added the floating dry dock back to the list of priority programs "to insure a properly constituted fleet" in the following year. The Navy's institutional attitude about what "a properly constituted fleet" consisted of had changed since the Washington Conference. Previous to that event the fleet and bases had been separate, although interrelated, elements of sea power. Now floating dry docks, the key component of the mobile bases the Navy planned to develop to offset the fortification clause, were part and parcel of the fleet. The original plans of 1922 had evolved into an array of linked programs that can properly be categorized as a balanced fleet. "Balanced" does not mean that the fleet was equally balanced as to numbers and costs for platforms. Rather it refers to a proper balance around the battleship with the necessary supporting and auxiliary classes needed to support the strategy of existing war plans. This balanced fleet was mandated by the treaty system itself and supported by a mobile basing plan that the Orange planners intended to use to sidestep the fortification clause. The annual building plans had always held a position of prominence in the serials of the General Board—at least prior to the 1930 London Conference.[47]

Events prior to the Washington Conference had already been pushing the Navy toward a more balanced fleet. World War I had emphasized the great utility of aircraft, submarines, and destroyers. The post-war Navy reflected this lesson learned by advocating a very healthy submarine construction program and by storing, instead of scrapping, its obsolete destroyers. Many of these destroyers, the famous "Four-Stackers," would be a central element in the expansion of the British Fleet under Lend-Lease in World War II. Also, the Battle of Jutland had taught the Navy the value that a scouting force of cruisers and fast battleships might have in a major naval engagement. Orange planning had also played a role in dividing the fleet according to function. The result was the division of the fleet into four parts—a battle fleet, scouting fleet, control fleet, and fleet base force.[48] The potential of air power caused Navy leaders to promote ubiquitous naval aviation for all the components of the fleet prior to November 1922. Finally, all of these trends remained consistent with the ideas of Mahan, who promoted an approach to naval design that focused not on one type to the exclusion of all others but instead "that in every class of naval vessel there should first of all, and first and last, throughout her design, be the recognition of her purpose in war."[49]

The second cruiser bill, passed after the collapse of the 1927 Geneva Conference, was the first concrete step toward a treaty navy that aimed at parity with Great Britain since Wilson's 1916 Navy Act. However, the real target was Japan. Despite the setback at Geneva, the governments of Japan, the United States, and Great Britain were still interested in naval limitation. Admiral Jones, now retired, was again dispatched to Geneva—this time for the Preparatory Commission meetings in 1928. The General Board had recently denounced an Anglo-French naval pact that established limits on the 8-inch-gun /ten thousand-ton cruisers but left smaller 6-inch-gun cruisers and below unrestricted. The Board recommended that using this pact as the basis for further limitation would make U.S. participation "fruitless."[50] However, the Anglo-French pact may have served as the motivation for a new means of measuring cruiser equivalencies nicknamed the "yardstick." The yardstick was a formula developed in the United States, possibly by Jones, to calculate "equivalent" cruiser tonnages. It used "tonnage, age, and gun power" as factors. A variation on this method was eventually adopted for use at the London Conference in 1930. It limited both classes of cruisers (8- and 6-inch guns) separately, but linked them by the yardstick. The details of how this formula was derived need not concern us here, but rather its results. The yardstick led to rapprochement on the cruiser issue between the United States and Great Britain that in turn led to the success of the London Naval Conference of 1930.[51]

The change in attitude can be attributed in part to the arrival in the White House of Herbert Hoover, one-time delegate to the original Washington Conference and advocate of worldwide disarmament. Hoover wanted to bring the British back to the table while accommodating the General Board's desire to build no fewer than twenty-three 8-inch-gun cruisers. To this end, Hoover presided over a marathon meeting of the entire General Board, the Secretaries of State and Navy, as well as the Chief of Naval Operations and the Undersecretary of the Navy in the White House on September 11, 1929. As a result, the administration decided to propose to limit the Navy to twenty-one 8-inch-gun (heavy) cruisers, ten older *Omaha*-class light cruisers, and five new 6-inch-gun (light) cruisers.[52] The fundamental problem was that the U.S. Navy desired to build as many heavy cruisers as possible for reasons already discussed. Lagging naval construction had brought the Navy to a crisis beyond any "cruiser gap" vis-à-vis the British or the Japanese.

As the discussion of the second "cruiser bill" illustrates, the Navy had real difficulty in getting anything built, not just cruisers.

The focus on cruisers generated by the treaty system offered the Navy a means to address another problem. This involved its dearth of naval aviation for Pacific operations.[53] At the suggestion of BuAer's Admiral Moffett, the Navy came to see the smaller class of cruisers in a new light. The still anemic pace of carrier construction in the United States due to costs had suggested to Moffett that perhaps cruisers were another way to get more aviation into the fleet. Moffett, in reference to light cruiser design, testified to the General Board, "I would say briefly that Aeronautics thinks that we should carry as many planes on cruisers as can be done without interfering with the proper mission of the ship." Moffett further recommended these new cruisers carry at least six aircraft to be launched by catapult.[54] The idea of putting as many airplanes as possible on cruisers was not new. A 1925 OpNav War Plans Division study had recommended installing flying decks on all U.S. light cruisers.[55] The Navy's plan "to build aircraft carriers at such a rate that the United States shall not fall behind treaty ratios" had been put on hold by the expectations generated by the treaty system.[56] The Navy reacted much in the same way as it had to other treaty-related setbacks, it compensated for a lack of capability by innovatively combining new technology with the operational requirements generated by the fortification clause. Long-range cruisers could be equipped with as many aircraft as possible within the terms of the treaty. The 1930 London Conference was an opportunity to implement this new dynamic for cruisers and aviation.

President Hoover wanted to ensure that the British and American positions would be closer than they had been at Geneva. Hoover went to great lengths to ensure that his delegation would compromise. In particular, Admiral Jones' presence on the delegation was balanced by the inclusion of the former President of the Naval War College Adm. William Pratt. Pratt was an advocate of the treaty system and had the confidence of the powerful Secretary of State and head of the delegation, Henry Stimson. Pratt effectively became the key senior Navy advisor for the conference. Jones's role in the conference was little more than that of a sightseer.[57]

In contrast to the Geneva Conference, the London Conference almost foundered on the rocks of the increasingly hard-line attitude of the Japanese over their inferior ratio to the American Fleet. They now had a much larger number of auxiliaries built than the 10-6 ratio proposed by the Americans. The Japanese delegation regarded 70 percent as the absolute minimum for any ratio given that they already had 80 percent equivalence to the United States in heavy cruisers. It was actually more than that since the Japanese heavy cruisers weighed about two to three thousand more tons than allowed by the treaty (this was not known at the time outside of Japanese naval circles).[58] At the eleventh hour a compromise was brokered by Sen. David Reed of the American delegation with Japan's Ambassador Matsudaira Tsuneo. Under this compromise the Japanese were given the 70 percent ratio for all auxiliaries except heavy cruisers (which would be 60 percent) and parity in submarines. In order to make this bitter pill easier to swallow for the Japanese, the Americans promised not to build three of their heavy cruisers until 1936, the year the London Treaty was to expire along with the original Washington Naval Treaty.[59] Perhaps as important in the long run, Great Britain, Japan, and the United States agreed to extend the Five Power capital ship building "holiday" to 1936.[60] This last element had been opposed from the start by the General Board but accorded well with President Hoover's commitment to both fiscal restraint and disarmament.[61] The General Board now had to delay its plans to begin building replacements for its battleships until after 1936. This in turn kept the focus of innovation on those classes, which could be built—including carriers, submarines, destroyers, deployable dry docks, and cruisers.

One important result of the Reed-Matsudaira compromise was that the relative importance of light cruisers increased. Unlike heavy cruisers, the United States could lay down these ships prior to 1936 without restriction. Not so well known was an interesting codicil with respect to these cruisers. Admiral Moffett, also a delegate at London, proposed language that would allow for "landing-on or flying-off" platforms to be attached to cruisers as long as "such vessel was not designed or adapted exclusively as an aircraft carrier." The Japanese, although suspicious of the plan, agreed to allow each party to build up to 25 percent of its total cruiser tonnage fitted with flying decks. Moffett's flying deck cruisers might ameliorate the problem of getting aircraft carriers to the fleet. Fleet aviation needs could be augmented by cruisers with up to 12 aircraft on board. Additionally, Fleet Problem IX in 1929 had emphasized a need for more aviation vessels. Limited to two large aircraft carriers, the fleet could lose most of its aviation support if either of these ships were sunk or damaged. All the eggs were in two baskets. Flying deck cruisers would spread the risk across the fleet and ensure that it was not so vulnerable to the loss of its aviation due to the sinking of the few precious carriers.[62]

The London Conference and the diplomats who negotiated there have been widely condemned for their failure to put in place commensurate political agreements as had been the case at Washington. Many observers also identify the London Treaty as the key turning point on Japan's path to aggressive militarism in the 1930s.[63] In the United States, the treaty was roundly criticized and the ratification process was not nearly as smooth as it had been for the Washington Treaties. It was publicly opposed by the General Board and most of the Navy. Many in the Navy blamed Admiral Pratt and when Pratt became CNO, his predecessor, Admiral Hughes, refused to shake hands at their change of command ceremony. However, some historians identify a new spirit of cooperation between the United States and

Great Britain as emerging from the London Conference.[64] Certainly the level of effort prior to the London conference on the part of the Hoover administration, as well as the installation of the anglophile Pratt as CNO after, did much to pave the way for increased U.S.–British amity.

After London the prospects for naval construction in the United States only worsened. This was due to two related factors. Foremost was the Great Depression that so bedeviled Hoover's presidency. Hoover adopted a policy of government "economy," which included across-the-board cuts, to include the Navy. Even before the stock market crash, shortly after he took office, Hoover had proposed "laying up" either one or both of the expensive-to-operate U.S. carriers *Lexington* and *Saratoga*.[65] Related to this was Hoover's sincere desire for disarmament through arms reductions. It was an article of faith for Hoover that arms reductions would lead to peace. Thus arms reductions and cuts in naval expenditures were mutually supporting policies. A memorandum from the office of the Secretary of the Navy to the president captured the mood and the lingering resentment over the London Treaty ratification fight: "The Navy is facing the most critical situation in its history. Upon the decision reached within the next few weeks depends the entire future of the naval defense [*sic*]. Formulation of a national naval policy in lieu of the one scrapped by the London Conference is irrevocably included in the building program now under consideration. . . . The bitterness of the fight against the London Treaty is not forgotten."[66]

Secretary of the Navy C. F. Adams directed the General Board to write a new *U.S. Naval Policy* based on the London Treaty. Instead, the Board submitted its recommendations that would have built the Navy up to allowed limits by the end of 1936 based on "the Limitation of Armaments of 1922 and 1930." In this response, the Board emphasized the now-limited role of the battleships in its considerations—modernization of these ships was already funded. The focus for naval construction switched to aircraft carriers, cruisers, and submarines. However, fiscal realities also played a negative role. The Navy dropped the large floating dry dock from its building plan so it could spend its limited funds on building warships and combat aircraft.[67]

The administration pressed on with its policies, further delaying construction of the new ships allowed under the London Treaty and authorized by Congress until after the League of Nations disarmament conference in Geneva in 1932.[68] The Navy's great fear, articulated as always by the General Board, was that the 1932 Geneva Conference would reduce the 35,000-ton limit for battleships codified at Washington and reaffirmed at London. The Board commissioned a special study by Cdr. E. M. Williams. Williams argued against any modification of the Washington capital ship limits. In the section of the study entitled "Arguments in support of reasons for the decision", primacy was given to the 35,000-ton displacement which in turn was due to the fortification clause: "A reduction in the unit size of capital ships would emphasize the value of a chain of naval bases to the disadvantage of the United States. Bases of operation are essential to the use of a navy." Lacking bases, the Navy needed these larger ships to obtain the endurance and range needed for War Plan Orange (see Chapter 5). After extensive further arguments, including a complete review of the terms of the fortification clause, the study continued, "The reduction of the size of ships, which tends to lessen the mobility of the ships . . . would operate in two ways to the advantage of Japan; for example, one, expose our own vital lines of communication, and two, secure the Japanese lines of communication from attack by the United States."[69]

Commander Williams' study emphasized the Board's approach to ship design and how it influenced innovation. Trade-offs of gun power, mobility, and armor would always have to be made. How to realize the most efficient balance between these factors was often determined during the General Board's hearings. Further, the influence of the fortification clause was also highlighted by the study's argument that any reduction in the size of capital ships and their armament would make U.S. Pacific strategy difficult if not impossible to execute in the absence of bases. Williams' study is among the first commissioned by the General Board during the interwar period to especially emphasize the critical role that the battleship's main battery would play in seizing bases as outlined in the Orange War Plans. Williams argued that the larger size of the battleships and their gun caliber were needed to both better protect U.S. lines of communications and for shore bombardment. Williams' predicted that the Japanese would install as large a caliber of shore gun as they could if war broke out. Williams' study must have been familiar to the United States Marine Corps since the General Board membership at this time still included the Commandant of the Marine Corps and he was on the distribution list for the Board's studies. Battleships were now being considered for use as an integral part of newer operational concepts, in this case how to use the fleet to support amphibious warfare with naval gunfire.[70]

World events set in motion in part by Japan's invasion and conquest of Manchuria undermined the 1932 Geneva Disarmament Conference. Hoover's sweeping proposals were meaningless in the face of Japan's aggression. The conference was eventually stalemated, not only due to Japanese defiance and intransigence but also over the increasingly disruptive issue of how to accommodate Germany, which was intent on repudiating the Treaty of Versailles and rearming.[71] Hoover's policies, especially his response to the economic depression, were repudiated at the polls with the election of Franklin D. Roosevelt, a former assistant secretary of the Navy.[72]

Hoover had sought the Board's advice for the next naval arms limitation conference in late 1932. The conference was tentatively scheduled for London in

1935—on the eve of the expiration of both the Washington and London Naval Treaties. This request was yet another attempt by Hoover to try to get the arms reduction movement and the treaty system back on track. The General Board did not deliver its opinion on the matter (Serial 1584) until after the election of Franklin D. Roosevelt (FDR). The timing of the delivery of the 33-page response—verbose even by General Board standards—suggests much. Perhaps the newly elected president would heed the Board in a way that his predecessor had not regarding the necessity of building the Navy to treaty limits. The serial reviewed in laborious detail the advent and course of the treaty system to 1932. It closed with the following ominous words, "Present preparedness must not be sacrificed to an illusory future readiness. National emergencies cannot be foreseen and must be met by existing forces." These words were for Roosevelt, not Hoover.[73] Hoover's commitment to disarmament could never be reconciled with the theory of sea power that had been advanced by Mahan and was still thoroughly embedded in the minds—the ideology, if you will—of the Navy's officer corps.

The new president did not disappoint the General Board's members, but neither did he meet all of their expectations. FDR had been an Assistant Secretary of the Navy during World War I and shared a common understanding of sea power with the Navy's leaders. He made no secret of his "love" for the Navy, even "unofficially" saying he "loved the United States Navy more than any other branch of our government" in his closing for a commencement speech at Annapolis.[74] However, Roosevelt was also serious about continuing the process of arms limitation. He believed he could have sea power and the political rewards that came with arms limitation by adopting a program to build the Navy to treaty limits. Nonetheless, he wrote the Navy no "blank checks" and held out great hopes for the next conference in London in 1935. Roosevelt also brought new attitudes to the equation: friendship, even partnership, with Great Britain, and a willingness to build ships.

Two key pieces of legislation in 1934 marked a turning point as the Japanese were withdrawing from the League of Nations: the implementation of the National Industrial Recovery Act (NIRA), which included $238 million for naval construction, and the first Vinson-Trammel Naval Bill.[75] The Vinson legislation committed Congress toward a program that would achieve the Navy's targets for numbers and classes of ships for a treaty navy over the next ten years. With the passage of NIRA Roosevelt was able to address two problems with one solution—providing much needed jobs and building a treaty navy that could both deter Japanese aggression in the Pacific or defend U.S. territory in the event of war. Ironically, Admiral Pratt, who had been most criticized within the service for his perceived part in stunting the treaty navy's growth, oversaw the initial period of this dramatic construction.[76] By the end of 1934 the handwriting was on the wall for the treaty system. The Japanese insisted that the basis for any future naval limitation be parity with the United States and Great Britain. When the United States refused to give this assurance in informal preconference discussions, the Japanese gave formal two-year notification, as specified in the Washington and London Treaties, of their repudiation of these treaties. No doubt the Vinson-Trammel Act and NIRA had also contributed to the Japanese resolve to leave the system.[77] Roosevelt hoped the Japanese could be convinced to rejoin the system at the upcoming London Conference, now scheduled for December 1935.

The Japanese did return to the table, but with their position fixed at naval parity. Great Britain was willing to appease the Japanese, but the U.S. delegation was firm and the Japanese withdrew from the conference early (January 1936). A treaty of naval limitation was signed between France, Great Britain, and the United States; however it offered numerous waivers for building in excess of limits by leaving intact the "escalator" functions of the original naval treaties.[78] If Japan exceeded tonnage or gun caliber limits, the other powers were authorized to exceed treaty limits. In 1936, the United States did precisely this in its design for the *North Carolina* class battleships. The 1935 London Treaty had limited new battleships to 14-inch guns. However, the president delayed approval of larger 16-inch guns allowed by the "escalator clause" until July 1937 when the Japanese confirmed their intent to build 16-inch-gun battleships. This action effectively ended U.S. participation in the treaty system.[79]

The period covered by the naval treaty system found the U.S. Navy often at loggerheads with the various administrations. Even Roosevelt desired—as did his predecessors—the reduction of arms expenditures through the mechanism of the treaty system. The Navy leaders of the period felt strongly that the nation needed to be prepared with the ships it had to meet national emergencies. This could not be done without a full-strength treaty fleet. The fundamental issue, which raised its ugly head over and over again, concerned the Navy's ability to exercise sea power given the constraints of the fortification clause of the Washington Naval Treaty. The Navy's arguments constantly returned to its lack of bases in the Pacific—sea power that had been "traded away" at Washington.

There was always a lingering hope in the Navy that somehow the fortification clause could be circumvented by traditional methods or that perhaps the treaty and the offending article would be abandoned. However, events proved that the United States would abide by the naval treaties signed at Washington and London until their repudiation by Japan effectively killed the treaty system. The hope that the expiration of the fortification clause would allow the Navy to address the problem traditionally was also in vain. Building up bases in the western Pacific (especially Guam) proved illusory in the face of domestic opposition to such funding. Also, political imperatives of long-standing duration, combined with an increasingly hostile Japanese government, limited the Roosevelt administration's options in taking concrete actions in improving the few western Pacific bases.

The fortification clause can be viewed as introducing an anomaly to the Navy's paradigm of sea power: How does one apply sea power without forward bases? The Navy innovated in response not only to come up with strategic solutions for an Orange War, but also to some degree it innovated in order to maintain the existing paradigm. It innovated within an existing system of thought. By so doing it significantly redefined the accepted boundaries of the paradigm itself, introducing the concepts of mobile basing, forward maintenance and support, and aviation facilities afloat. But the essentials of the paradigm, that sea power depends on bases to support it, protection and development of merchant commerce, and the need for a fleet to defend or attack remained fundamentally the same at the end of the treaty period as it had at its beginning.

* * *

The status quo conferred by the fortification clause had in effect become the norm, even after the constraints of the clause no longer had the force of law. The intersection of Navy leaders' conception of sea power with the constraining effects imposed upon it by the treaties forced the Navy to seek solutions elsewhere. A combination of technological and operational innovation became the means for pursuing strategic goals in the midst of new political and material conditions after 1922. Stymied in its efforts to effect change by modifying the treaty system, the General Board sought new solutions. Although the battleship remained the primary measure of naval power, its importance was less than that prior to the Washington Conference.[80] The battleship still represented power and dominated naval minds, but Navy officers had wrestled with its limitation and diminution for so long that a fundamental change in attitude had occurred, one that would only be fully exposed by the debacle at Pearl Harbor. The demise of the battleship at Pearl Harbor was the occasion not for a revolutionary but rather evolutionary change in attitude.

The Navy's institutional change in attitude toward the role of aviation was less complicated. Certainly the importance of sea-basing aviation was boosted by the fortification clause. The Navy knew it needed naval air power, especially after the battleship bombing demonstrations of the 1920s. It also knew that air power from land bases would not be available in any great numbers in the opening phases of an Orange War. The Navy's cultural and institutional attitude, which recently had viewed naval aviation as at best a novelty and at worst a waste of funds, now saw aviation as essential.[81] This was in no small measure due to the treaty system's devaluation of the battleship and the fortification clause's constraints on basing. With the mandate islands in their possession, the Japanese could quickly build airbases, whereas the U.S. Navy must bring its air power to bear—in defense or offense—aboard its ships.

The change occasioned by the limitation on bases, and therefore logistics, as an element of sea power was more radical. It was a direct result of the fortification clause. Although there were pre-1922 trends toward a longer-ranging Navy, the first treaty—like a catalyst—pushed the Navy toward a new way of thinking about sea power in the absence of forward bases. As with aviation, the Navy would bring what it required with it, to include prepackaged bases and floating dry docks. This is a challenge that the U.S. Navy still wrestles with on occasion such as the United States' loss of its Philippine naval and air bases shortly after the end of the Cold War.

The influence of these changed attitudes, especially among the membership of the General Board, which translated policy into programs, contributed to the formation of a balanced fleet that leveraged technology as much as possible.[82] It was reflected in the Navy building policies promulgated by the General Board beginning in 1923. These reflected the master plan crafted by the General Board after the Washington Treaty was signed. This plan envisioned a treaty navy that would have 5-3 fleet superiority over the Japanese Navy at the earliest possible date. This fleet was centered on a limited number of modernized battleships. The battleships would be supported by long-range cruisers and submarines, also designed to operate without permanent bases. Additionally, this fleet was to have the maximum amount of aviation to protect it and to offset its lack of land-based air support.

The next three chapters address three program areas of the treaty fleet that were rich in innovation—battleship modernization, naval aviation, and the mobile base project. These initiatives and programs subtly changed the way the Navy thought about overseas bases, the Navy's third element of sea power. Navy leaders and planners began to conceive of supporting the fleet upon the advent of war without preexisting bases. Amphibious warfare, mobile bases, underway replenishment, submarine warfare, and embarked air power were concepts that matured in the Second World War, but they were conceived and embedded in the Navy's collective consciousness during the interwar period.

The General Board's members implicitly accepted the OpNav planners' innovative mobile base plan as essential to the support of the treaty fleet. This fleet would fight and establish interim bases on its way across the Pacific, to either relieve or recapture the Philippines. This support was reflected in the building plan by the presence of large floating dry docks in the annual Board serials for building policy year after year, despite a lack of funding. The mature plan for a treaty fleet was best represented by the 1929 building program (crafted in 1927) and involved the construction of a completely different sort of fleet—one that could range far and apply sea power in the absence of preexisting overseas bases. This was the design for the fleet that defeated the Japanese Navy in 1942 in the Coral Sea, at Midway, and in the seas around Guadalcanal—mostly without its battleships.[83] In this manner, the treaty system unintentionally kept the United States from building a less capable fleet than it otherwise might have.

Chapter 5

Battleship Modernization

The battleship is still the backbone of the fleet and the bulwark of the Nation's sea defense, and will so remain so long as safe navigation of the sea for purposes of trade or transportation is vital to success in war.
                                            The Joint Board, August 1921

The sea officer must use the tools which science creates. Of these the most powerful and the most fully known is the battleship.
                        Prize Essay, United States Naval Institute, 1938

The battleship remained the centerpiece of all fleets after World War I, despite its varied performance during that conflict. Battleships did not win the war at the Dardanelles or at Jutland—but neither did they lose it. After the war new developments diminished, but did not replace, the battleship as the centerpiece of the U.S. Navy (or any other fleet). Navy doctrine and treaty constraints influenced the decisions of the General Board as it deliberated what to do with the battleships left to it. Battleship modernization by the U.S. Navy from 1922 to 1937 shows how naval treaties of the interwar period influenced one aspect of innovation at the tactical level.

The "party line" for the interwar U.S. Navy—as for all the other major navies—was that the battleship remained the final arbiter of naval power. By the end of the treaty system in 1937 this remained the mantra of the U.S. Navy. The treaty system relied on capital ship dominance as the basis for success. However, by 1937 the Navy had hedged its bets, in part because it was forced to by the treaties, but also because the attitudes of naval officers had gradually changed. This was especially true of those officers who were to lead the Navy during World War II. They had "grown up" in the treaty fleet. Interestingly, when not actually with the fleet, much of their creativity had been focused on building and planning for everything but battleships. The plans for the battleship's use were structured mostly around how to get to the fight. Ironically, these plans included very little in the way of new concepts for the use of the battleship. Often, as in the 1924 War Plan Orange, the actual use of the battleship in an engagement was simply sidestepped with the phrase, "These plans in no way limit the initiative and decision-making of the operational commander."[1] They would be used as they had in the past, to engage and defeat the enemy's battle fleet. Also, these same war plans forecast battleships as the platform of decision, but not as the initial—or even intermediate—weapon of decision. Instead, the battleship shared the stage with aviation, aircraft carriers, cruisers, submarines, destroyers, and amphibious warfare. The initial stages of a war might see the aircraft carrier in an all arms task force that might include only a few of the fastest battleships, or even the unglamorous floating dry docks, as more important to setting the stage for the final Tsushima-like victory of the last phase. Battleships were precious, too precious to risk early in the conflict. It is ironic that the attack on Pearl Harbor reinforced the U.S. Navy's predilection to hold back its main battleship force and instead rely on the other ship classes of the balanced, and mostly new, treaty fleet.

Battleship modernization emphasizes that innovation after the Washington Conference initially proceeded along traditional lines. The battleship was traditional in the sense that it was still the centerpiece of the U.S. Fleet and as such improvements to the battleship received first priority in the budget for funding. Secondly, battleship innovation was gradual. No major leaps in battleship design occurred during the period. The improvements—turret elevation, propulsion modernization, and defensive improvements—were changes to existing designs. The navy did conceptually plan for the day when it would be able to build new classes of battleships, but ironically—mostly because of the London Treaty—these designs had the potential to be less powerful than the most modern treaty battleships such as *Tennessee* and *West Virginia*. For example, the *North Carolina*-class battleships only received approval for their 16-inch guns in 1937, after almost being fitted with 14-inch guns instead. The innovations associated with the modernization of these vessels tended to focus at the tactical level, although money was also spent to ensure that the battleships would be able to perform their operational role outlined in War Plan Orange. Improvements tended to focus on existing design trends in the navy that supported existing doctrine.

Operationally, the Navy had always placed a premium on fuel efficiency and long radius of action because of the vast expanses of the Pacific and Atlantic oceans. The modernization program for the propulsion systems of the battleships was not a radical change, but rather an evolutionary, and again gradual, type of innovation. However, here too, the fortification clause channeled designs along certain paths. Fuel storage, type, and corresponding propulsion systems as design considerations were highlighted because of their impact on the operational range and staying power of these ships. Use of the term "modernize" comes from the language that the interwar Navy consistently used to characterize its reconstruction of battleships throughout the period. What the Navy meant by the term can be found in the broader language of the 1922 *U.S. Naval Policy* expressing the Navy's aims: "To make superiority of armament in their class an end in view in the design of *all* fighting ships. To provide for great radius of action in *all* classes of fighting ships." (emphasis mine).[2] Although gun armament was important to all the navies of the interwar period—it was practically a cult in Japan—radius of action was less so. Of the great naval powers, the United States was the most affected by the treaty restrictions in this area because of the fortification clause. Thus, radius of action became as important as gun caliber and armor in the hierarchy of design criteria.

Armament and radius of action became the pillars for the Navy's warship design in general and for battleship modernization especially. Modernization also has another temporal meaning, improvements made to ships to better allow them to operate in the modern (current) environment after World War I. This was reflected in a section of the Washington Naval Treaty known as the Reconstruction Clause. This clause allowed for the reconstruction (or modernization) of battleships in order to allow them to operate in the contemporary naval environment that now included air and submerged threats (mines and submarines).[3] Battleship modernization in the U.S. Navy also highlights the organizational aspects of innovation in the interwar Navy. As usual, the General Board served as the nexus for this effort. The U.S. program for battleships highlights the role of the treaty system and the difficulty of determining what could and could not be done under some of its more ambiguous language. This example also shows the occasionally innovative ways in which Navy leaders interpreted—or even circumvented—the treaty system to attain the maximum performance allowable for their battleships.

* * *

The Navy's plan to create an efficient treaty fleet involved the intersection of technological advancements, new strategic and material conditions conferred by the treaty system, and novel operational approaches about how to operate at extreme distances without secure or available bases. This plan included:

1. the modernization of the battleships retained under the treaty,
2. the development and construction of all allowable aircraft carriers, cruisers, and submarines and,

3. the construction of a mobile base force with an extensive supporting logistics train (the "fleet train") for use upon the outbreak of war.

Naval aviation was envisioned as being ubiquitous throughout the fleet and augmented to the maximum extent allowable from shore. The program with the highest initial funding priority in this overall plan centered on the eighteen battleships that the Washington Naval Treaty allowed the U.S. Navy to retain.[4]

It was no accident the Navy concentrated its efforts on the battleships first. Prior to the Washington Treaty, the Navy saw capital ships as the decisive element of the fleet. With the implementation of a capital ship building holiday the focus shifted from building new battleships to modernizing those battleships retained under the terms of the treaty. As predicted by naval historian and Com. Dudley Knox, naval competition moved into the sphere of "efficiency." What Knox meant by efficiency was that navies, the U.S. Navy especially, would focus on getting as much performance from their systems and personnel as possible. In Japan, the "inferior" status conferred by the treaty contributed to redoubled efforts in the realm of training. The Japanese hoped to ameliorate what they saw as a material disadvantage to the U.S. Navy by relentlessly practicing night-fighting tactics. In the U.S. Navy a different approach was taken by focusing on a number of areas, not just superiority in battle. Systems would have to synergize with one another. For example, increased fuel storage could be accompanied by designs that optimized storage and fuel transfer for damage control and anti-torpedo protection. In this way radius, defense, and damage control could all be improved at the same time. This was what the Navy meant by efficiency.

The fortification clause played a key role in the modernization of these ships. Coal-fired propulsions systems were converted to fuel oil propulsion (for the six oldest ships) and turboelectric drive systems in use on the most modern ships were targeted for installation on the older ships. These conversions gave these ships greater cruising radius, better engine performance, enhanced crew comfort for longer periods at sea, and ease of refueling. The modernization of these ships was dictated by what the Navy thought was allowable under the terms of Treaty. Tactical doctrine also played a role in the adaptation of the airplane in extending the usable range of battleships' main gun batteries. In turn, modification of the battleship turret to allow it greater gun elevation was seen by the U.S. Navy as essential to achieving the greater ranges airplane spotting theoretically made possible. This was especially true given that the Treaty prevented the employment of larger guns or the significant modification of the layout of existing battleship main batteries.[5] The General Board immediately began hearings on the plans for the battleship program beginning in April 1922 with the ink barely dry on the Washington Naval Treaty.

The final modernization program for the battleships had three main elements: improved propulsion systems, improved gun turrets, and treaty-allowed weight increases of up to three thousand tons to allow the major powers to improve the defenses of battleships "against air and submarine attack." Admiral Pratt, who had been a technical advisor for the American delegation at the Washington Conference, was present during the adoption of this last element of the program. He later testified to the General Board that this proposal was the idea of the British and its intent was to allow the battleships to keep pace with threats by aircraft and submarines, which were not limited under the terms of the Treaty.[6]

After World War I, U.S. naval officers expected the battleship to remain the decisive element in naval warfare. However, these same leaders were puzzled. The battleship, although preeminent, was not as dominant as it had been thought to be prior to the war. Submarines, mines, torpedoes, and airplanes posed new challenges to the mainstay of naval power in all three dimensions—air, surface, and undersea. The intersection of several of these "new technologies," such as the combination of the torpedo with the airplane or the submarine and the mine, posed especially vexing challenges. It is easy to forget or downplay the dizzying changes that the naval leadership of all the major powers, often conservative by nature, faced during and after World War I.[7] Technological changes occurring during World War I had traumatized and then mobilized naval leaders of that generation. These men were not trying to recover their lost world as much as they were trying to keep pace with change.

After World War I things did not get any easier for Navy leaders. The first challenge to the battleship was the end of the Great War itself. With the end of this "war to end all wars" naval officers, especially in the United States, faced unavoidable cutbacks to the unprecedented naval building program initiated by President Woodrow Wilson and Secretary Josephus Daniels in 1916. This program would have given the United States the most powerful fleet in the world. The U.S. public understandably felt that now that victory had been achieved there was no longer a requirement for such a large Navy. The 1916 Navy Act stimulated a costly postwar naval race. This race, along with tense relations with Japan, had in large part led to the Washington Naval Conference. U.S. admirals were already uncertain about the utility of the battle cruiser based on the British experience at Jutland and had decided to center their fleet on what later came to be called "fast battleships." These designs were scuttled by the Washington Conference. One indicator of the Navy's attitude toward battle cruisers was that the General Board in 1921 had already investigated and tentatively decided on converting half of the new battle cruiser hulls into aircraft carriers. There was no intention to replace the converted hulls with newer battle cruisers.[8]

The other more famous challenge came with the bombing tests by Billy Mitchell and the Army Air Corps, especially the sinking of the German dreadnought *Ostfriesland* in 1921. At the time the Navy rightly pointed out the artificiality of this demonstration: the ship could not maneuver, it had no anti-aircraft armament, no personnel were aboard to conduct damage control, and the weather had to be perfect. However, Navy leaders took this demonstration to heart in private even as they contested its results publicly. Despite Cdr. Van Keuren's (the naval observer for the tests) assessment that the German ship was in poor material condition and "slowly taking water all the time," Navy leaders took the sinking of the *Ostfriesland* by Mitchell's bombers quite seriously. They concluded that it had been the "near misses" that had contributed most to the demise of the already-sinking ship. The final words of Van Keuren's report stated, "In other words, since 'The shots that miss are the shots that count' in this new form of warfare, we must see that the least possible number of shots are fired by hostile airplanes and that those that are fired go very wide of their mark." This conclusion was to have a dramatic influence on battleship modernization and design under the treaty system.[9]

Theodore Roosevelt Jr. led a discussion that removed a bit more luster from the battleship during the Washington Conference's deliberations on submarines. As discussed in Chapter 3, the U.S. delegation opposed the abolition of the submarine, principally due to the Navy's desire to use submarines to ameliorate the effects of the fortification clause. However, Roosevelt and others also argued against outlawing of submarines due to their record in the late war against the battleship. They included the following reasoning in their minutes: "It may be noted that in the Admiralty return of August, 1919, to the House of Commons entitled 'Navy Losses,' it was shown that out of a total of 13 British battleships lost in the World War, 5 were sunk by submarines, over 38 percent. . . . Its effectiveness as a mine layer against men-of-war is again shown by quoting from the same Admiralty return that 5 battleships out of 13 were sunk by mines. In view of the number and activity of the submarine mine layers and of the circumstances under which the vessels [battleships] were lost, it is fair to presume that two or three of those [5] battleships were sunk by mines planted by submarines."[10]

In the light of this evidence, the committee then went on to conclude that "the United States needs submarines to defend itself."[11] Submarines could sink battleships and must therefore be retained.

The ultimate challenge to the continued viability of the battleship came not from the air or from under the sea, but from the arms limitation movement. After the seemingly indecisive clash at Jutland, and the minimal employment of the U.S. battle fleet during the war, many civilians questioned the need for any of these ships—especially if airplanes could sink them. One observer at the Washington Conference quipped that "[Secretary of State Charles Evans Hughes] sank in thirty-five minutes more [capital] ships than all the admirals of the world have sunk in a cycle of centuries."[12] No other ship class was as limited and constrained by the Washington Treaty as was the battleship. Previously a nation with a dominant navy, such as Great Britain, could simply out-build challengers and structure its policy (as Britain did prior to World War I) in order to be reasonably assured of command of the sea in wartime. After Washington this was no longer possible with the 5–5–3 ratio. Not only could nations no longer build as many behemoths as their purses would permit but they were prevented from building any of them at all for at least ten years (later lengthened to almost 15 years). Finally, the very size of the ship and its guns were limited. These limits, when combined with the building holiday, had a profound influence on the design and improvement of battleships in the interwar period. Other classes of ships could be designed with various hull sizes to accommodate various power plants and a plethora of the sort of innovative opportunities that designing and building a ship from the keel up provided. Not so the battleship. Any improvements had to be made without altering the existing hull, armor, and armament of those ships retained under the Treaty.

U.S. Naval leaders could not afford to ignore what they saw as threats to the centerpiece of their fleet and, ultimately, their strategy. Seeing themselves foremost as professionals, they felt obligated to come up with solutions. Now they had to come up with solutions within the framework of the Washington Naval Treaty. First among these solutions, as reflected in the Building and Maintenance Policy of 1922, was "to keep all retained capital ships modernized as far as treaty terms permit, and good practice justifies."[13] The rules for the replacement, scrapping, and alteration of capital ships and aircraft carriers under the Washington Naval Treaty were contained in its replacement section. This is where the language for the naval holiday resided, and which prohibited any replacement tonnage under the treaty from being "laid down until ten years from November 12, 1921." However, in the same section of the Treaty, the following language allowed for alterations to existing capital ships:

> d. No retained capital ships or aircraft carriers shall be reconstructed except for the purpose of providing means of defense against air and submarine attack, and subject to the following rules:
>
> The Contracting Powers may, for that purpose, equip existing tonnage with bulge or blister or anti-air attack deck protection, providing the increase of displacement thus effected does not exceed three thousand tons (3,048 metric tons) displacement for each ship. No alterations in side armor, in calibre, number of general type of mounting of main armament shall be permitted except:

1. In the case of France and Italy, which countries within the limits allowed for bulge may increase their armor protection and calibre of the guns now carried on their existing capital ships so as not to exceed 16 inches (406 millimeters); and
2. The British Empire shall be permitted to complete, in the case of the *Renown*, the alterations to armor that have already been commenced but temporarily suspended.[14]

This clause came to be known as the Reconstruction Clause. According to this clause, the major powers were allowed for increases of no more than three thousand tons weight when modifying or reconstructing their capital ships and carriers. The carriers had been included because they were being converted from several battle cruisers scheduled to be scrapped under the Treaty by Japan and the United States. This allowance reflects directly the impact of the threats to the battleship of the torpedo, submarine, mines, and aircraft. The Treaty's framers allowed battleship designers some latitude in increasing the survivability of the existing ships to meet these threats over the next ten years. This was the clause under which the United States would press ahead with a substantial program of battleship modernization under the Treaty.

Propulsion Modernization and Target Tests

The Navy's first priority in the modernization of its battleships was propulsion. This focus came as much from the bureaus as it did from the General Board's internal discussions. The four major bureaus concerned with modernization of the battleships—BuAer, BuOrd, BuEng, and BuC&R—forwarded a joint letter to the Secretary of the Navy with their general recommendations on April 4, 1922. In this letter they emphasized that "It is understood that the General Board considers it part of Naval Policy to keep all retained capital ships modernized as far as the terms of the Treaty permit." Their first recommendation stressed making use of "materials and machinery intended for those vessels [the scrapped ships], especially such machinery as is now in the course of construction, as much of this material and machinery could well be devoted to modernizing existing capital ships." From this recommendation the Bureau chiefs pointed out that the limitation of three thousand tons in the reconstruction clause "requires a clear understanding." Accordingly, they recommended that the General Board convene hearings to determine what the Treaty permitted and then to decide on a plan of action. They followed this general advice with detailed areas for consideration by each of the Bureaus. Primacy of place in their list belonged to engineering and they recommended modernization of "Boilers and Propelling Machinery," "Conversions of coal burners [six battleships] to oil burners," and "Electric Light Plant and Communication" improvements. Of key importance was the intersection of engineering improvements with internal structural improvements in order to take advantage of the allowance for three thousand more tons of protective reconstruction (see Appendix 3). The letter stressed that conversion of the propulsion systems not only led to better fuel efficiency and power, but conversion to oil and internal reconfiguration of fuel storage compartments would also improve torpedo protection. Unlike coal, fuel could be transferred more easily. Greater numbers of watertight (or fuel-tight) compartments contribute to a ship's survivability as it allows damage to be sealed off and localized. The addition and rearrangement of compartments, and what they were filled with, enhanced the survivability of a ship after a below-waterline hit. Flooding in particular was easier to control and fuel could be moved around instead of counter-flooding with seawater in order to better stabilize a ship after it had been hit. Finally, the bureaus stated:

> Owing to the scrapping of so many capital ships now under construction there will be available a great deal of modern machinery and boilers—both finished and unfinished—which could be used for, or adapted to, the purpose of giving modern economical propelling machinery to those vessels which are to be retained. . . . This proposed machinery is of the type that permits the best form of torpedo protection. Moreover the installation of this machinery will not only result in very greatly improved military characteristics, such as a homogenous fleet with similar maneuvering characteristics *and increased cruising radius*, but it will also result in a very considerable saving in money expended for fuel and maintenance costs are bound to be reduced. (emphasis mine)[15]

The letter then emphasized that this course of action was not what the bureaus would typically recommend, preferring instead to construct new ships altogether. This recommendation was a result of Treaty limitation on the battleship, "which demands that the retained capital ships shall be modernized as far as the Treaty permits regardless of the fact that the cost will be high and the result never as good as when built new." The joint letter from the bureaus became the basis for the propulsion and anti-torpedo modernization for the battleships of the interwar period.[16]

In summary, the major elements of the modernization program recommended by the four bureaus consisted of conversion of all battleships to oil-burners, modernization of electric generation systems, and the use of as much of the existing machinery off the scrapped ships as possible to perform these functions. The conversion of the coal-burning battleship *Texas*, for example, would result in almost double her cruising endurance from a range of 7,600 nm to over 15,000 nm.[17] BuAer separately recommended modifying the battleship gun turrets to allow greater gun elevations that would give more range. Admiral Moffett believed greater gun efficiencies at greater range were now realizable due to improved aircraft spotting techniques. It seemed only natural to him that the new technological capabilities of the aircraft must be taken advantage of, Treaty or not. The operational and tactical fruits of the resulting program would be a longer-ranging, more fuel efficient, harder-hitting, and homogenous battle fleet.[18]

The General Board received the joint letter from the bureaus on April 11, 1922 and hearings were scheduled soon after for April 17–18. These hearings saw the General Board follow precisely the path recommended by the bureaus. The first hearing addressed what was allowed, or not covered, under the Treaty. The Board called upon its new member, Admiral Pratt, fresh from the Washington Conference as a technical advisor, "to determine the import, so far as may be" possible, of the reconstruction clause. After reading the clause verbatim, Captain Schofield (who was conducting the hearing) introduced Admiral Pratt who was then questioned by the Board.[19] The General Board really wanted Pratt's advice about what the Treaty would or would not allow in reconstructing the battleships. The Board first wanted to know if the internal structure of the ship could be changed without violating the Treaty. They also wanted to know Pratt's opinion regarding the use of fuel oil boilers and turbo-electric drives from the ships that were to be scrapped for installation on the older ships, especially the six battleships that were still coal-fired. Pratt suggested that propulsion system changes could be made under the Treaty as long as weight limitations were not exceeded. However, when queried about replacing existing armor with new composites of greater strength, Pratt thought "it would be violating the intent of the Treaty as far as I could ascertain from the discussions that went on." The issue of turret reconstruction, interestingly, did not come up, despite its having been mentioned in earlier correspondence to the board by Admiral Moffett of BuAer.[20]

It was from this early meeting that Pratt became the de facto "expert" on the spirit and intent of the original Treaty. He remained so over his long association with the Board, first as a committee member, then as president of the Naval War College, and finally as CNO. Later, he would be joined in this role by Adm. Hilary Jones. The subsequent hearings of the Board make clear that Pratt was regarded, and consulted, as the expert on the Washington Conference and Treaty. Pratt's role emphasizes, again, the Board's characteristics of openness and flexibility in gathering advice, expertise, and opinions. Other members of the General Board often disagreed with Pratt's interpretations of the Treaty, but he was usually consulted anyway. In any case, Pratt, as a member of the General Board, always had the right to offer his opinion whether it was in conformity with the consensus of the Board or not.

The Board's hearing on April 18 proceeded, on the basis of Pratt's testimony, from the assumption that the machinery and propulsion changes proposed by BuEng were allowable. The bulk of the hearing was devoted to the assignment of new fuel oil boilers from the scrapped ships to the six remaining coal-burning battleships in the fleet: *New York*, *Texas*, *Florida*, *Utah*, *Arkansas*, and *Wyoming*. Also discussed were the turbo-electric drives from the scrapped vessels for use on the older ships. The representative from BuEng told the Board that these changes would increase fuel efficiency, torpedo protection, and cruising radius. BuEng and BuC&R both recommended that all the battleships—except for the five newest ones—receive substantial modification inside the hull that would both improve fuel storage and torpedo protection. Admiral Harry Huse, one of the members of the Board, emphasized that "in the absence of fortified stations in the Far East the increase in radius is of very great importance." Huse's comment led to further discussions between the Senior Member, Admiral William Rodgers, and the bureau representatives about other ways to use this opportunity to increase fuel performance, and thus radius of action. They came up with the idea of switching the newly acquired geared cruising turbines from the scrapped ships for those retained ships that did not have them yet. These turbines had better fuel efficiency at lower speeds and thus would assist in giving the ships a longer radius of action.[21]

Later that month, the General Board summarized these hearings and recommended the following program for battleship modernization:

a. that the retained coal burning battleships be converted to oil burning as soon as practicable;
b. that sufficient boilers, now built or building for capital ships that are to be scrapped, be retained to meet re-boilering demands and to reboiler six coal burning battleships should these be converted to oil burning;
c. that the plans and estimates made of time and cost of conversion of all retained coal burning battleships to oil burning include increased protection against torpedo and air attack;
d. that the electric machinery of six battleships that are to be scrapped be completed and held in readiness for installation in battleships already completed until detailed plans and estimates are available for determining the desirability and practicability of such installation;
e. that detailed plans and estimates of time and cost be prepared without delay for modernizing the *Nevada* class; and
f. that electric drive be not substituted at present for the machinery installations in capital ships.[22]

The Board endorsed this plan again in its next serial on Navy building policy. The only recommendation not adopted by the Board, possibly due to its questionable legality under the Treaty, was the conversion to "electric drive." The conversion to electric-drive turbines would end up taking place on those battleships modernized after the coal-burners (the *Nevada* class).[23]

These hearings, and the Board's subsequent recommendations, emphasize how the Treaty channeled innovation along certain paths. By including a reconstruction clause, the Treaty opened the door for the Navy to ameliorate the damage done by the fortification clause through the mechanism of modernization of its battleships. Instead of a limited program of improvements to provide torpedo and air protection, the General Board (stimulated by the bureaus), used the occasion to propose a sweeping conversion of its six oldest battleships to fuel oil propulsion. The remaining ships would also be modernized and given updated fire-control systems, enhanced deck protection, and both internal and external torpedo protection. This program also allowed the Navy to maximize fuel storage and rearrange and modify the compartmentalization inside all the ships to gain as much protection against torpedo and other underwater explosions (like bomb near-misses) as possible. As it turned out, many of these changes added no weight, especially to the older coal-burners. The Navy, intent on using every ton, ended up adding the blisters allowed by the Treaty to the exterior of the hull for these ships, which added additional space for fuel storage and in some cases actually translated into higher speeds. As seen, conversion approximately doubled the radius of action of the old coal-burning battleships to an endurance of over 15,000 nm.[24]

The second phase of battleship modernization encompassed the modification of the seven oldest fuel oil-burning battleships. These ships would benefit from the electric drives saved from the scrapped ships since none were ever installed on the oldest ships. The final phase addressed "the Big Five," which were the most modern battleships. These ships could also be reconstructed under the Treaty. It was hoped that by the time their turn came lessons learned during the modernization of the others could be applied to them, especially those involving underwater protection.[25]

Interestingly, the proposals for propulsion modernization were also enhanced by another element in the Treaty that had not been foreseen by the Treaty's framers. The Navy decided to use the occasion of the scrapping of its most modern capital ships in order to test the efficacy of its new anti-torpedo designs for the conversion of the retained battleships. The Treaty allowed the Navy to use of one of the ships scheduled for scrapping each year as a target for testing and research. Given the experience of the *Ostfriesland*, the Navy was keen on taking advantage of this clause.[26] The battleship *Washington*'s use as an experimental target was immediately proposed. She was the Navy's most modern battleship scheduled for scrapping under the Treaty. Someone on the Board had broached this issue because use of the *Washington* for experimentation had been adopted by April 22, 1922 as the Board's official position. Admiral Rodgers stated that he knew "of no reason why we cannot destroy the *Washington* by target practice in the period set by the treaty and be in every respect within the terms of the [T]reaty."[27] The *Washington* was not the first ship to be disposed of in this way. In 1923, the older *Virginia* and *New Jersey* were bombed and sunk by the Army Air Corps. Service feeling over these tests tended to increase, rather than weaken, the Navy's resolve to strengthen the battleship as well as beef up its own fleet aviation.[28] The next ships disposed of were the battleships *North Dakota* and *South Carolina*. These ships were altered with the proposed conversion designs in mind, especially the *South Carolina*. Their use as targets provided valuable data to test the blister design for the coal burners and their deck protection designs. Valuable information was also gathered on shock protection for gun turrets.[29]

President Coolidge approved *Washington*'s use as a target in August 1924. The Navy Secretary had emphasized to the president that the tests were experimental and "in the public" interest since the results would be studied "by a board of naval experts" for use in the modernization program.[30] The subsequent tests showed the Navy designs for survivability of her class (the *Tennessee*) to be extremely sound. The *Washington* tests, which would not have been held if not for the Treaty, established in the Navy's mind the soundness of its continued support for the battleship. These tests did much to defuse the Navy's concern over near misses that had resulted from the *Ostfriesland* and other level bombing tests, while providing the Navy with valuable underwater data to enhance torpedo protection. *Washington* was eventually sunk by 14-inch gunfire from *Texas*, which further reinforced the Navy's commitment to improve deck protection and to attempt to lengthen the range of its guns as permitted by the Treaty. Attitudes about the central role of the battleship changed little inside the institutional Navy. Most Navy leaders redoubled their efforts to retain it as the coin of naval power.[31] However, in maintaining this stance, leaders changed their views in other areas, especially as regard the threat of air power and submarines. These nontraditional platforms gained importance in the minds of the interwar naval leadership, in part due to their dedication to the idea of the survivable, big-gun battleship.[32]

Turret Modernization

Tactics, not just new threats from above and below the sea, also dictated the course of the General Board hearings on battleship modernization. At the time of the Washington Conference, the U.S. Navy had identified engagement (or gun) range and the speed of its battle fleet as areas of deficiency. The Navy had come to these conclusions based on its study of the Battle of Jutland. Also, Navy tacticians focused on the failure of the Royal Navy at Jutland to bring its superior firepower (especially its ability to out-range the German guns) and speed to bear on the German Fleet, and in so doing, inflict more damage. These tacticians eventually decided that the proximate cause of this failure was "the inadequate coordination and communication of the British Forces." However, the Navy first had to rectify, as much as it could under the Treaty, its deficiencies in speed and gun range at the same time as it focused on improving command and control.[33] The problem faced by the Navy was how to improve the gun range, speed, and command and control of its battleships in order to prevail in an engagement with another (presumably Japanese) fleet.

The Treaty had eliminated many of the traditional solutions that might be employed to resolve these problems such as building newer hulls with more powerful propulsion plants and bigger, longer-ranging guns. The naval holiday prohibited building new battleships altogether. Also, the 35,000-ton maximum weight and 16-inch gun were design limits tied to the fifteen-year life of the Treaty.[34] However, the naval officers of the era, including those in the United States, suspected the battleship might be even further reduced and restricted. Some could even foresee a building moratorium that lasted a lot longer. Additionally, naval officers knew there were proposals to eliminate the battleship altogether, just as there were proposals to abolish the submarine and the airplane from warfare. These fears were justified. The building holiday was extended for the life of the Washington Naval Treaty, the battleship was scaled back in numbers and gun size at London, and proposals were forwarded to abolish the submarine, airplane, and battleship at London in 1930 and again at Geneva in 1932.[35]

These concerns had the impact of creating a very anxious environment during the interwar period about what to build. However, for the time being, the battleship remained a legal, and dominant, weapon of war. Navy leaders were aware that great changes could take place, even though a sort of status quo had been achieved. Because of this atmosphere, due mostly to the treaty system, changes and innovations that would not ordinarily have been readily contemplated gained wider acceptance while at the same time Navy officers sought to ease their anxiety through support of the battleship modernization program. Thus the traditional and the nontraditional resided side by side. In the United States, the strategic anxiety produced by the Treaty was perhaps more acute than in the other powers. This was in no small part due to the fortification clause. The verbiage in the war plans, War College problems, and estimates of the situation practically scream the U.S. Navy's anxiety over its perceived inferior position in the Western Pacific due to Article XIX. One citation from the 1924 Orange Plan exemplifies the anxiety produced by the Treaty: "The lack of suitable means to repair underwater damage in the Western Pacific is at present *the most serious material difficulty that will be encountered in the prosecution of an Orange War.* The deficiency of fresh water is next in seriousness. These defects will have a most far reaching effect in attaining victory and, if not remedied by zero day, are very liable to prolong the war to a grave extent. (emphasis original)[36]

The Treaty did not cause an instant revolution in naval thinking, but it did create an atmosphere of anxiety and uncertainty that slowly changed the way naval officers thought about their strategic problems in the Pacific.[37]

Navy tactics prior to the Treaty dictated engaging the enemy at the longest ranges possible. By doing this, the Navy would conform to Mahan's tactical maxim to "strike the first blow" while remaining out of range of the enemy. To this end, the Navy had focused on building battleships armed with 16-inch guns.[38] At the time of the Washington Treaty, the United States had a slight superior hand in this type over the British and the Japanese—they each only had two battleships with 16-inch guns while the United States was allowed to keep three of this type. The Treaty froze this superiority in place. However, the bulk of the American Fleet was slower than its Japanese and British counterparts.[39] In addition, because of its desire to engage at extreme ranges, the Navy adopted an "all or none" armor policy for its battleships that maximized the armor around key components of the ship. On the plus side, this design innovation protected the most vital mission areas of the ship while saving weight to allow increased speed, fuel efficiency, and fuel storage. However, the lack of armor over other less critical areas of the ship made Navy battleships vulnerable at shorter ranges to smaller caliber guns with higher rates of fire. Once the smaller ships closed to within range the unarmored areas above the waterline, especially command and control and fire control stations, were more vulnerable.[40] Because of this combination of factors, the U.S. Navy was committed to a tactical doctrine of extreme-range engagements with the 16-inch gun as the preferred gun to achieve it. However, the problem remained as to how to get the remaining ships that were armed with 14-inch guns for the next ten years the same advantages of range that the few 16-inchers had. The Treaty prevented conversion of existing batteries from 14 to 16 inches.

Two technological innovations intersected to provide a partial solution. The first problem in fighting at extreme range was how to accurately spot the fall of shot at the most distant ranges in order to adjust and maximize the gunfire. Hit accuracy tended to be very low at these long ranges. Anything that might improve long-range accuracy would increase an enemy fleet's vulnerability in the long-range killing zone. The emergence of airplanes with reliable radio offered one part of the solution. They were, figuratively speaking, elevated, remote "crow's nests" that could extend the gunnery range of the fleet in good weather. Aviation technology intersected with a device developed during World War I known as the Ford

Rangekeeper, a computer that could track two moving objects with respect to each other. The combination of aerial spotting and the Ford Rangekeeper might be termed a minor revolution in naval affairs, because it combined targeting information collected beyond visual sight by aircraft with an integrated computing system that gave the U.S. Navy a potentially winning edge in naval gunnery.[41]

The confluence of aircraft spotting with range finding had come about in part as a result of testimony before the General Board in 1919 by Lt. Cdr. Kenneth Whiting. Whiting reported that spotting experiments with the battleship *Texas* had led to a 200 percent improvement over the advanced spotting technique that used kite balloons. Kite balloons, in turn, were a big improvement over the standard gun-spotting/range finding techniques that relied on spotters elevated in the battleship's armored mast.[42] The range at which the tests were performed was also critical—20,000 yards, right at the limit of *Texas'* guns (*Texas* had 14-inch guns). Of more importance was the Board's reaction to this report. The BuOrd representative was "very enthusiastic" and wanted to "see a spotting plane on each [battle]ship." More remarkable was the reaction of the second-most senior member of the Board, Adm. Albert Winterhalter. During a discussion of shore-basing spotting airplanes versus having seaplanes that could accompany the fleet far out to sea, Whiting emphasized his preference for sea-based planes. Winterhalter added, "Something that will land in the sea and will do its other work too. If it *can* be made so, it must be *made* so." (emphasis original)[43] Winterhalter's emphasized words indicate that the Board was most anxious to get aviation out to sea, especially if it meant longer ranges and the ability for higher hit percentages for its battleships. In this case, the innovation was of the most common kind—modification and improvement to the existing ships and doctrine using new technology. The innovation's impact was manifested at the tactical level. Improved gunnery performance demonstrated by the use of aerial spotting and new computing devices led to a desire to improve the performance of all the battleships in order to obtain the maximum ranges possible. The problem was that the Treaty prevented just the sorts of improvements (increased guns sizes and radically new turret designs) that might remedy the shorter ranges of the Navy's older battleships. Or did it?

The irrepressible Admiral Moffett had already proposed that the way around this conundrum was to increase the elevation of the guns. Moffett had recommended "alteration of turret gun mounts to permit firing at maximum range: This to enable us to take advantage of our development of spotting gun fire from airplanes, which it is believed is much further advanced in this service than in any other Navy."[44] This suggestion was not initially well-received. Someone from the Board had handwritten in the margins next to this proposal "No" with a question mark next to it, perhaps indicating that on initial inspection it appeared a violation of the Reconstruction Clause language but in retrospect maybe not. Clearly Mof-

Vice Adm. William V. Pratt in January 28 as Commander, Battleship Divisions, Battle Fleet. Pratt had previously served as a delegate at the Washington Conference, on the General Board, and as president of the Naval War College. He later commanded the entire fleet and as the fifth chief of naval operations was perhaps the most innovative CNO of the interwar period. He was extremely open to new ideas such as the flying deck cruiser and was one of the first officers to propose independent carrier task forces. He was an early proponent of the Naval Treaty Limitation System. (NH 77489)

Adm. Hilary P. Jones on USS *Columbia*. Jones has often been regarded as a man "who saw the world through a porthole" by naval historians. However, he was among the most skilled naval diplomats of the interwar period and an expert on the naval arms limitation system. He headed the U.S. delegation to the failed Geneva Conference that resulted in the first substantial naval building program since the Washington Conference. Although he and Pratt often disagreed, they had a collegial working relationship and both believed in the importance of naval aviation and power projection. (NH 83088)

Adm. Mark L. Bristol with staff officers on board USS *Pittsburgh*, 1928. Bristol, Commander in Chief of the Asiatic fleet (CINCAF), was one of the Navy's diplomat admirals having earned some notoriety as the first U.S. ambassador to the Republic of Turkey. He later brought his considerable diplomatic experience, both in Turkey and the Far East, to his assignment as chairman of the executive committee of the General Board during the crucial period of the early 1930s. (NH 84500)

A classic shot of the General Board of the Navy in 1932 in the old Army-Navy building (now the old Executive Office Building). This shot is contemporaneous with the meetings and hearings over the issue of design of the Flying Deck Cruiser. Note the civilian dress. Those seated are (left to right): Rear Adm. Mark L. Bristol, Rear Adm. Charles B. McVay Jr., Capt. John W. Greenslade, Cdr. Theodore S. Wilkinson (secretary), Rear Adm. Jehu V. Chase, and Capt. Cyrus W. Cole. Standing are (left to right): Lt. Col. Lewis C. Lucas, USMC (Ret.), and Cdr. Edgar M. Williams. Note also the portrait of Admiral of the Navy George Dewey, first president of the General Board, on the wall to the left. (NH 50175)

Assistant Secretary of the Navy Henry L. Roosevelt, a distant cousin of the president, with the Fleet Flag officers in 1934. A number of key innovators are present in this shot. Sitting (left to right): **Vice Adm. Walton R. Sexton**, Commander, Battleships, Battle Force; Vice Adm. Frank H. Brumby, Commander, Scouting Force; Admiral Sellers; Assistant Secretary Roosevelt; **Adm. Joseph M. Reeves**, Commander, Battle Force; **Vice Adm. Harris Laning**, Commander, Cruisers, Scouting Force; and Rear Adm. Thomas T. Craven, Commander, Battleship Division ONE. Standing (left to right): Rear Adm. William S. Pye, chief of staff to Commander, Scouting Force; Rear Adm. Adolphus Andrews, chief of staff to Commander, Battle Force; **Rear Adm. Frederick J. Horne**, Commander, Base Force; **Rear Adm. Edward C. Kalbfus**, Commander, Destroyers, Battle Force; Rear Adm. Charles P. Snyder, chief of staff to Commander in Chief, U.S. Fleet; Rear Adm. John Halligan, Commander, Aircraft, Battle Force; Rear Adm. Henry E. Lackey, Commander, Cruiser Division FOUR; **Rear Adm. Adolphus E. Watson**, Commander, Destroyers, Scouting Force; Rear Adm. Charles R. Train; Rear Adm. Manley H. Simons, chief of staff to Commander, Cruisers, Scouting Force; and Brig. Gen. Charles H. Lyman, Commander, Fleet Marine Force. Key General Board members are in bold. (NH 76413)

One of the larger sectional floating dry docks in use in World War II (YF-326). These docks were based on the original designs first proposed during General Board hearings in the mid-1920s. (File number TR-8765)

A smaller sectional floating dry dock in transit during World War II. This may also consist of just a couple sections from a larger dock. The sectional docks were built in a modular fashion and the sections could be towed (or moved under their own power) to a central location for assembly into larger docks. (File number 84138)

This photograph is a wonderful depiction of the "Treaty Fleet" at the advent of war. YFD-2 is in the foreground being positioned by fleet tugs. On the far left behind Ford Island is *Yorktown* (CV-5), to the right of Yorktown are two destroyers, the aviation tender *Wright* (AV-1), two light cruisers (CL), and in the left center is an unidentified destroyer-sized seaplane tender (AVD). The battleships are probably out of port conducting target practice with their guns. (Courtesy of the Naval Historical Center)

fett thought that modifying turrets was allowable under the Treaty. After all, the turret was not to be "replaced" but only "altered." This proposal remained unexamined through the first two General Board hearings on capital ship modernization. Guns were only briefly mentioned by Pratt during the April 17 hearings and turret housings were not discussed at all.[45]

However, the idea did not take long to catch on. By December turret alteration had become one of the major elements of the modernization program. Typically, the proposal was given a thorough review during the hearing process. Also, it appears that Moffett had continued to lobby on behalf of the idea since it was linked to integration of naval aviation with the main fleet. In this case it appears Moffett heavily lobbied the Bureau of Ordnance. By October 1922 BuOrd had forwarded disturbing reports on the inadequacy of the guns for the majority of the battle fleet vis-à-vis Great Britain. The first priority of these hearings, according to Admiral Huse, was "Increasing the elevation of the guns to get greater range."[46] The key question, as with internal modifications for propulsion, was whether the Navy could make turret alterations under the Treaty.

Adm. Charles B. McVay of BuOrd was given the floor in order to provide estimates as to how much range could be attained by altering the gun ports in order to allow greater gun elevation, which would in turn give these ships greater range. McVay argued that by increasing elevation, for example, on the battleships *Utah* and *Florida* to 30 degrees they could almost double these ships' gun ranges to 33,500 yards. Even the newer *Arizona* class could benefit by these modifications with gun ranges up to 34,000 yards. McVay stressed that BuOrd's designs covered increases for all thirteen of the older battleships. Only after McVay made his very promising presentation was the issue of the legality of this modification raised. Adm. Joseph Strauss of the General Board objected to McVay's use of the term "mount" as he outlined replacing older hydraulic mounts with electrical mounts. The Treaty stated that "No alterations in side armor, in calibre, number or general type of *mounting* of main armament shall be permitted . . . " (emphasis mine).[47] With this the discussion, in contrast to that of the previous April, became contentious:

> Admiral Strauss: I should say that a vessel that had a hydraulic mount and shifted to an electrical mount would shift from one type to another, for instance.
>
> Admiral McVay: No. I think that [an alteration of turret control] would modify it unless you made an absolute change.
>
> Admiral Strauss: "Mounting" is used in the prohibition.

> Admiral Huse: I think that is a little dangerous to put [it] on record as the opinion of the General Board.
>
> Captain Stocker: The French version [of the Treaty] is a little better. This [English] is, "No alterations 'in number or general type of mounting.'"
>
> [At this point a further discussion with regard to the restrictions imposed by the Treaty was had, not for the record].

The transcript's closing parenthetical remark indicates that part of the discussion about the impact of the Treaty was serious enough to cause deletion from the record. Note also the Board's use of the French translation of the Treaty because its wording favored the alteration that was being proposed. In this case the issue of whether the changes proposed by McVay could be categorized as "general" alterations of the mounts was critical; if they were, then they would be prohibited. The Board adopted a very narrow attitude about what "general" meant. The Board's interpretation of "general" mounting was that the caliber, number of the guns, guns per turret, and the location of the turrets were fixed. For example, a battleship with nine 14-inch guns mounted in three turrets, with three guns per turret, two turrets forward, and one turret aft could not be "reconstructed" to place an additional gun in any of the turrets or perhaps move the aft turret forward. Everything else, to include elevation angles and turret control, was deemed outside the Treaty and could therefore be modified as part of the modernization program. When the official transcript resumes, it picks up with a discussion of the cost for the various turret modifications.[48]

The Treaty came up again later in the hearings. Admiral Coontz, the CNO, had been silent for much of the hearing. Coontz's comments summarized the considerations that the Board was forced to make in rendering both design and policy decisions. It is also an example of the extraordinary influence of the General Board in these matters:

> The General Board will undoubtedly shortly make a preliminary settlement of what it thinks it wants to do. Whether it wants to tackle old battleships first or some of the later ones; whether it is going to attempt the entire amount of these modernizations; whether it thinks that certain ones should be done quickly in the line of the gun elevations and the line of the deck [AA protection]; then the question of modernization regarding oil-burning and so on. It has a number of very knotty problems. The question of time; the fact that the treaty may be signed for twenty years.

. . . then the question of money must come in. If it is to be a vast sum, whether we could get it all or whether we had better get ready to do it at a later date. The question of how long it will take the British [to finish the capital ships and their modernization programs].[49]

It was not uncommon for someone of the status of Coontz to hold back, and then summarize the issues the Board must address. However, the bulk of the discussion about the legality of alterations to improve gun elevation under the treaty must have occurred during the discussion omitted from the transcript. The final reference to turret elevation came from Admiral Huse, "Some of this work is imperative. I don't think anyone will object to increase in range of guns." Evidently the issue of legality had been resolved as far as the Board was concerned. In December, the Board made modification of the turrets to achieve greater gun elevation for the thirteen oldest battleships the number one priority of the battleship modernization program.[50]

The program for turret modification had the complete support of Secretary of the Navy Denby. He had mentioned 6.5 million dollars set aside for turret modification as having "a very good chance" of gaining Congressional funding in a letter to the Navy League in early 1923.[51] However, this proposal immediately ran into opposition from other quarters. On February 26, 1923 the British Admiralty lodged a formal protest stating that the gun elevation modification violated the terms of the Treaty. This was enough to cause the administration to pull the plug on this program, despite funds having been allocated by Congress to support it. The money for turret modification was also removed from the General Board's 1923 building policy recommendation for the 1926 fiscal year.[52] In the meantime, Denby had the General Board prepare a legal review of the turret issue so that the administration would have all the facts. This review, not surprisingly, found that none of the modifications proposed violated the Treaty since they did not alter "the general type of mounting." Captain Schofield, the author of this review, closed the serial by suggesting that the British protest was never "stated officially through the proper channels" anyway. Schofield's opinion represented the consensus of the Board that the whole issue was moot and that the U.S. could modify turret elevations legally under the Treaty—formal British protest or not. This review made extensive use of the French translation of the Treaty in rendering its opinion.[53]

Opposition to turret modification now reverted back to Congress. Once Secretary Denby halted the funding due to the British protest, funding for the program was not easily restored. In 1924 the new Chairman of the Executive Committee of the General Board, Adm. Hilary Jones, emphasized the legality of the modifications and the need to restore the funding that had been removed from Congressional authorizations in 1923. Jones lobbied for an emergency authorization to allow funds to be used immediately for the 1925 building and modernization program.[54] However, Jones' appeal was in vain. Turret alterations were not included in the battleship modernization bill approved that May.[55] The Navy continued to request the funds each year but by the time they were finally approved (1926) turret modifications for the newer battleships were deemed of greater priority than modifications to the six oldest ships. This was primarily due to the possible impending retirement of the oldest of these ships once the capital ship holiday expired and the Navy was again allowed to build replacement shipping from the keel up. Why make modifications to ships that may only have four or five years of useful service left? These ships retained their original turret elevation and limited range for the rest of their service lives. Three of them were still in action during World War II in the Atlantic (none had been at Pearl Harbor). This left the seven battleships of the *Nevada* and *New Mexico* classes as the number one modernization priority in the U.S. Navy (henceforth referred to as the intermediate battleships).[56]

The issue of gun elevation was finally resolved, although not to the Navy's complete satisfaction, in 1926. Early that year, former Secretary of State Hughes, who had been consulted by Admiral Pratt about treaty violations, had written Congressman Butler on the House Naval Affairs committee assuring him that the Japanese did not consider the elevation alterations a violation of the Treaty. The General Board then proposed that money be authorized to alter the gun turret elevations of only the seven intermediate battleships in its building recommendation for that year. Subsequently, Navy Secretary Curtis Wilbur asked Congress to fund these alterations beginning in fiscal year 1928. Congress authorized the funds but the modifications, along with the others allowed under the Treaty, were not begun until fiscal 1930.[57] The British, however, continued to complain. Navy Secretary Curtis Wilbur wrote to Ambassador A. B. Houghton to have him relay to the British that "[former] Secretary of State Hughes . . . Secretary Kellogg . . . the naval officers [Rodgers and Pratt] who participated in the [Washington] conference, and the Congress that enacted the law for the elevation of guns, all believe that the treaty does not prohibit increased gun elevation." Case closed.[58]

In 1928, Admiral Moffett joined in the efforts to get turret elevation altered on the seven intermediate battleships as soon as possible. Moffett used the latest gunfire results from the 1927 Battle Fleet reports to stress the need for increasing the gun range on the older battleships: "Long range battle practice was fired at ranges never before attempted in our Navy, the five long-range ships firing at approximately 30,000 yards. The *West Virginia*'s performance at this practice indicates beyond all question the vital necessity for an increase in the maximum elevation of turret guns of our short-range battleships."

The report that Moffett cited went on to emphasize "the practicality of engaging an enemy effectively beyond the horizon" using aerial spotting. Although the report implied that the purpose was to engage ships beyond visual range, or hidden behind smoke or land, the implications for indirect fire support by the battleships were also highlighted by Moffett.[59]

The impact of the treaty system on turret modification was substantial. In the first place this program may not have taken place at all had the Treaty not been signed. Its genesis as a means to bring the U.S. fleet up to modern standards coincided with the Navy's efforts to ensure its battle fleet could engage adversaries at the longest ranges possible. These long ranges were made possible by the emergence of both aviation and advanced fire control technology. In the long run, however, the Treaty indirectly slowed the Navy's efforts to implement these innovative improvements and prevented them being made at all on the Navy's six oldest battleships. By the time the legality of modifying the turrets on the older ships was established, budget priorities for battleship reconstruction had shifted to newer ships with more service life. However, if part of the fleet had to be limited these should certainly have been the ships that the Navy chose to leave deficient. The subsequent 1930 and 1936 Treaties at London had little impact on this program, since they only constrained new-construction battleships after the expiration of the building holiday.

Finally, the General Board was not willing to simply accept at face value what the Treaty stipulated. As one Admiral had commented about the London Treaty, "that is usually the joker in these treaties. One paragraph says you can do a thing and another paragraph says you can't do it."[60] As often as not the members of the General Board defaulted to a permissive interpretation of the treaties, one that allowed them a way around a particular constraint. This was a habit of mind familiar to the Navy, since in previous years it had been domestic factors, not international ones that had to be surmounted in building what they thought was the appropriate fleet needed for the security of U.S. interests. What was new was the context attributable directly to the treaty system. They now had to wade into less familiar international and diplomatic waters. Although the members tended to be conservative and parochial, they were required by events to assume a different, albeit expanded, role in foreign affairs.

This new role required a different set of skills than required of previous generations of Navy officers. Compromise did not come easy to these old sea dogs who often thought that concessions were forced upon them. For a younger generation of officers—the Thomas Kinkaids and the R. K. Turners—their participation in the treaty system was more fundamentally formative. For both mid-grade and senior officers their participation in the treaty system changed, and sometimes broadened, their outlook. Ultimately the civilian administrations, and even former Secretary of State Hughes, found the General Board's arguments about the plan for battleship modernization, including turret elevation, compelling. Also, despite the sustained objections of the British, the United States prevailed in its interpretation of the Treaty. This interpretation had a positive impact on innovation at the tactical and operational levels of war in the case of battleship modernization.

The Other Elements of U.S. Battleship Modernization

Gun elevation and propulsion systems were the major elements of the battleship modernization program. They were both results and, to a degree, victims of the treaty system. Both can be seen as responses to the strategic and tactical impact of the Washington Treaty on the Navy. The fortification clause in particular impelled the Navy to radically improve the propulsion systems of these ships. The building holiday that had frozen battleship tonnage affected modernization at the tactical level as well. The Navy still saw itself inferior in terms of the engagement ranges for the majority of its ships and therefore sought to use the reconstruction clause of the Treaty to allow it to enhance gun range by modifying its battleships' turrets.

Concurrent with the gun and propulsion programs were those modifications made to improve submarine, anti-aircraft, and deck protection. The Washington Naval Treaty specifically allowed these improvements and implementation was limited only by the budget process. They were often delayed because the Navy's desire to accomplish as much of the reconstruction concurrently in order to save money. Thus, the delays associated with the approval of the propulsion system and turret elevation modernization proposals were transmitted to these programs as well. The anti-submarine programs, in particular, intersected with the propulsion designs in improving the survivability of the battleships should they sustain below-waterline damage. By the time the mid-1930s arrived, a new president was in office and the very latest improvements to the newest battleships (the "Big Five") could be contemplated. As it turned out, the price tag for the improvements the Navy wanted to make to these ships was too high for both Congress and the president. By the time the United States could afford to modernize them the treaty system was defunct. Some of these ships were still undergoing modernization when the Japanese attacked Pearl Harbor (e.g., the *Colorado*).

During the treaty period the purpose of the battleships themselves continued to come under scrutiny. The imperatives of the Orange War Plan strategy reflected in the OpNav/NWC problems had caused the Navy to consider battleships in a different light. Strategic Problem IV from the 1926 War Plans Division had outlined the traditional role of the battleship in an Orange War: "The exposure of her [Blue] battleships and battle cruisers to damage and destruction cannot be justified by any gain that might be accomplished by their use in the early stages of a Philippine campaign. These facts, therefore, point to their [capital ships] unsuitability as part of Orange effort at the present time. Such type are indispensable in the later stages of the Orange campaign of attrition that is sure to be waged."[61]

In other words, battleships were too valuable to be used early in the Orange campaign. What might they do prior to the engagement in the "later stages of the Orange campaign"? By 1931, with the treaty fleet at perhaps its lowest ebb in terms of morale and equipment, Cdr. E. M. Williams outlined a new role for these ships for the General Board:

> A reduction in the unit displacement of the capital ship and a reduction in its offensive power would increase the relative superiority of guns on shore over those afloat. . . . This paper has emphasized the lack of bases on the part of the United States and the importance of bases. No more need be said on that particular point. It has been shown that it may be necessary to seize bases. While these bases may not now be heavily fortified, it would be a relatively simple matter for an enemy, upon the outbreak of war, to emplace heavy artillery before the United States fleet could be placed in a position to move against them. Should such events come to pass the fleet could not seize the bases if the guns of the bases outranged those of the fleet.[62]

Williams argued for the retention of the 35,000-ton size and maximum gun caliber for U.S. battleships for the purpose of naval gunfire support for seizing advanced bases. In this specific case the treaty system encouraged innovative operational thinking for using battleships in support of amphibious warfare. Although the Navy's commitment to these ideas faded to some degree prior to Pearl Harbor, it would bloom again after that fateful event. Williams' report was prescient, for shore bombardment was precisely what the treaty battleships ended up doing for most of World War II.

* * *

The Washington Naval Treaty was the proximate cause of the modernization of the U.S. battleship fleet during the life of the treaty system. Without this Treaty, the United States would probably have instead spent its money, as the bureaus had pointed out in their joint letter prior to the 1922 hearings, on new battleships rather than modernizing old ones. Also, the Treaty set the strategic and tactical context within which these programs were implemented. Without the fortification clause, the design of any new battleships would probably have been different given the premium on tonnage conferred by this clause. One can imagine that Navy designers would have focused less intently on the radius of action requirements. When the General Board began preparing for the 1935 Conference in London it emphasized in one of its studies that it could accept no reduction in the 35,000-ton maximum weight of replacement battleships because "the Navy may and probably will be required to conduct a campaign in . . . distant areas."[63]

The course of battleship modernization during the treaty period illustrates how the attitudes of Navy officers, both young and old, changed in response to a new strategic dynamic. Historians have commented on how the Navy came to play a larger role in the foreign affairs of the United States in the late nineteenth and twentieth centuries.[64] In the twentieth century this role was not solely due to the naval limitation treaties, but certainly the Navy's role in foreign policy formulation and implementation was enhanced by them. Secondly, innovation in the Navy after the Washington Conference tended to proceed along traditional lines. The battleship was the centerpiece of the Navy and as such was the first program to receive substantial funding for improvements—not new construction for carriers and other auxiliaries. These innovations tended to be at the tactical level, although much money was also spent to ensure that the battleships would be able to perform their operational role as outlined in the Orange War Plans rewritten after the Washington Naval Conference. These improvements tended to focus on already existing design trends in the Navy that supported operational and tactical doctrine. The treaty system, especially the building holiday, channeled and influenced what could be done with the battleship more than any other class. On offense, this was reflected in the increased ranges of guns and, on defense, in improved protection allowed by the Treaty against submarine and air threats.

The modernization of the battleship also highlights the intersection of technology with tactics. Airplanes enabled new tactical possibilities while these same possibilities encouraged the continued development of naval aviation. At the nadir of naval construction during the period (1931), the impact of the treaty system, especially the continued threat to the size of the battleships, caused Navy leaders to consider new operational uses for the battleships such as shore bombardment in support of amphibious warfare. The Navy had always placed a premium on fuel efficiency and long radius of action because of the vast expanses of the Pacific and Atlantic.

However, the fortification clause of the Washington Naval Treaty channeled and bounded these design considerations in the bureaus, the War Plans Division, and the General Board. By limiting the tonnage of these ships, and what could be done to reconstruct them until the building holiday ended, the Treaty forced the Navy to consider ever more carefully design tradeoffs that limited the range of action of these ships. In limiting what could be done with armor, the Treaty caused the Navy to emphasize radius of action and gun range in its modernization plans. There was only so much weight allowed and the Navy must spend it wisely and still have a fleet that could operate in vast expanse of the Pacific until the logistics problems of the status quo could be remedied, that is, until forward bases were established.

To a greater degree the fortification clause also reinforced the Navy's strategic inclination to let the other parts of the fleet do the initial fighting in an Orange War while it held its battleship force back for the decisive naval battle implied by the Orange Plan's "thrusting" course of action. Viewed in this way, Navy leaders, by switching to the more methodical "step-by-step" course of action did not eliminate the decisive battle from their thinking, but they certainly put it on the backburner for the sorts of operations they envisaged for the first part of any Pacific war.[65] More and more they came to see a Pacific war as one based on attrition, not decisive battle.

Finally, the fate of the battleship as a component in the strategic plans of the United States reflects a great irony. Navy planners and leaders contemplated a war whose opening phases saw the battleship playing only a secondary role. The battleship was the strategic reserve to be used *after* the auxiliary naval forces. These auxiliary forces centered on aircraft carriers and would be used to develop the strategic situation. On the other hand, the battleship might be used as gunfire support for amphibious warfare to support, especially after 1935, the cautious "step-by-step" course of action that came to hold prominence within the interwar Navy. As it turned out, those battleships that were to be the core of the reserve, the newest of the treaty battleships, were mostly destroyed at Pearl Harbor. Meanwhile those that remained were of the older type and soon assumed the role of amphibious support bombardment, many never leaving the Atlantic Fleet. The newest, faster battleships—built after the treaty system collapsed but conceptualized during the treaty period—came to fulfill the role of escorts for the fast carrier groups.[66] All of these uses had been contemplated prior to World War II. The Navy had been thinking for some time, in part due to the treaties, how to fight the enemy fleet prior to getting bases and prior to committing its battle line. When the day did come that this battle line was committed, at Surigao Strait during the Battle of Leyte Gulf, it was something of a surprise because the role of the battleship had changed so much.

Chapter 6

Naval Aviation and the Fortification Clause

Aircraft will settle next war. Don't care how many surface vessels we build. The more the other nations build, the less money they will have for aircraft. Friday March 21 proposed to Captain VanKeuren he get up a design for a 6-inch gun cruiser that would have a landing deck for aircraft.

Rear Adm. W. A. Moffett Jr., March 24, 1930
at the London Naval Conference

If battleships represented the course of innovation in the U.S. Navy along more traditional paths, then innovation in the area of naval aviation represents equally the traditional and non-traditional. Traditional refers to the use of aviation to support the battle fleet. Aerial spotting was a technological improvement at the tactical level—a supporting function to the offensive use of the fleet. On the defense it was envisioned that the indigenous aircraft supporting the battle line, including those from the carriers, would defend the battleships from the enemy's aircraft—especially land-based bombers, which after the *Ostfriesland* sinking were regarded as the primary airborne threat. This role was similar to those missions assigned to cruisers, destroyers, and submarines. Just as destroyers might protect against the torpedo, airplanes would defend against the airplane. Finally, the aircraft would perform reconnaissance—similar to the role of auxiliary warships like cruisers—a role that was both offensive and defensive.[1]

The development of naval aviation as an operational means to apply sea power was not limited to the aircraft carrier. The leadership of the U.S. Navy promoted the development of a variety of aviation programs and initiatives in order to solve the perceived problem of a lack of sufficient aviation in the fleet, especially in the

Far East. In particular, the development of aircraft tenders for the Asiatic Fleet and the later flying deck cruiser proposal emphasize how Navy leaders used a variety of solutions and different ship classes to attempt to solve strategic problems as they designed the treaty fleet. The use of aircraft in an offensive role independent of the big guns was a mission that was initially secondary to the supporting functions of fire control, air defense, and reconnaissance. Other than aerial spotting, the roles envisaged for aviation reflect innovation at the operational level.

If we accept the idea that the traditional naval approach to battle used a group of battleships as the force of decision, then the idea that naval aviation could be used in the absence of the battle line in the Far East can be categorized as a non-traditional application of naval power. After the Washington Conference, naval aviation was viewed as a necessary component in support of a traditionally structured naval force that would aid in the defense of the Philippines, while at the same time whittling down the battle line of the enemy. Once within range of the enemy it would eliminate the threat from the air and then switch roles to aid in fire control and in harassing attacks to break the cohesion of the enemy's formation.

It was no accident that as the size and character of battleships remained static (due to the building holiday), and with the distinct possibility of a war without forward bases, that naval air power became more important. As these views evolved, naval aviation's offensive potential came to the fore but naval aviation could not be employed without runways. In the Far East it would have to be based on carriers in large numbers and in lesser numbers on other ships. It must be supported from the sea because the fortification clause denied the United States the means to support naval aviation ashore. It may seem obvious to us today that the balance of naval aviation must be aboard ships at sea but this truism was not so obvious at the time of the Washington Conference. In fact, it was aggressively challenged by land-based air power advocates such as Billy Mitchell—a challenge only put to rest by the Morrow Board in 1925.[2] However, for Navy leaders the fortification clause clarified this truth—in a war in the western Pacific, naval aviation must be operationally based at sea. How could this be done in an era of naval limitation, disarmament, and government economizing? This was the challenge Navy leaders faced. They addressed it by investigating various ways to get aviation to sea—on submarines and destroyers (possibly), battleships, tenders, carriers, regular cruisers, and maybe even on a new class of ship called the flying deck cruiser.

* * *

Navy leaders planned to use aviation as one means to ameliorate the impact of the fortification clause on naval strategy in the Pacific. The War Plan Orange of 1924 clearly stated that: "[the strategy] involves particularly the offensive use of naval forces and *the use of Army and Navy air power overwhelmingly*; resting on a naval main body, of strength superior to the entire Navy of Japan"[3] (emphasis mine). Navy leaders initially thought that airplanes and submarines could be more easily based in the status quo infrastructure of the Philippines—however, airplanes and submarines were not limited by the Treaty. This assumption soon proved problematic, especially with regard to aviation.

Typically, Admiral Moffett spearheaded the efforts to apply an aviation solution to the challenges posed by the Treaty in the outfitting and deployment of aviation in support of the Manila-based Asiatic Fleet (AF). Senior officers such as Admiral Sims and Captain Schofield supported BuAer's efforts. Schofield came by this view in a roundabout way. He had initially hoped that the army would help the navy around the constraints of the fortification clause. Schofield addressed the Treaty's impact on the army in a 1923 speech at the Army War College on the topic of the Washington Conference. He emphasized how the lack of bases between Hawaii and Manila negatively affected the army's and the navy's ability to defend the Philippines: "How long would it take the army and the navy to get command of the sea in the Far East, because the Navy alone cannot get that command of the sea. It must have the army to hold its shore bases and to preserve those bases in every respect free from enemy land and air attack."

Schofield noted that the Treaty did not prevent basing "unlimited numbers" of submarines and aircraft in the Philippines and that he hoped " . . . no occasion will be lost to increase gradually the strength of these two arms in the Philippines."[4] At this time Schofield was still a supporter of the "thrusting" Orange course of action versus the more cautious "step-by-step" approach. In the thrusting scenario, the navy would rapidly deploy the fleet to help defend the Philippines in order to retain the U.S. naval base at Cavite near Manila. However, the army's ability to help the navy soon ran afoul of army budget cuts, complaints by the Japanese that the United States was violating the Treaty, and a realization by Navy planners that the Philippines would probably be lost in any case. The U.S. bastion in the Far East would have to be recaptured rather than defended.[5]

The 1924 Orange War Plan and the Naval War College's Strategic Problem IV both emphasized the importance of ship-based aviation in the absence of land-based air.[6] Navy leaders were unsure as to what this aviation force should look like, but they knew they wanted one and were willing to experiment. In the case of the U.S. Asiatic Fleet, several solutions were proposed. Few of them were adopted. Land-based naval air was presumed to be prohibited by the fortification clause. An innovative plan was crafted by the General Board, BuAer, OpNav War Plans Division, and the Asiatic Fleet to build or convert existing ships into an airplane tender fleet that could serve as floating bases for the naval air force of the Far East. This plan eventually proved to be logistically flawed and fiscally unrealistic. Over time, Navy leaders realized that the bulk of the naval aviation for a

Western Pacific campaign would have to come from the main fleet as long as the fortification clause remained in force.

In 1922, the General Board advocated that the converted battle cruisers *Lexington* and *Saratoga* be completed "now" in order to best determine, through a process of experimentation, what designs to use for the construction of the remaining aircraft carriers allowed under the Treaty.[7] However, the slow introduction of these aircraft carriers to the fleet caused the Navy to investigate other ways to take advantage of the operational promise of naval aviation. Navy efforts focused on a variety of ways to integrate aviation into fleet operations. These efforts show that the Navy not only spread the risks in preparing for a future war across the surface and subsurface forces, but across aviation programs as well. Moffett and the General Board advocated a balanced naval air concept, one that even contemplated the combination of the submarine with the airplane.[8]

In addition to combat planes embarked on warships, the navy considered the possibility of using tenders, also called "air depot ships," in an effort to increase the naval air component afloat. These ships eventually became known as seaplane tenders (designation AV), but this was only because the Navy felt constrained to equip the Asiatic Fleet with seaplanes since it was thought that basing naval air power ashore was prohibited by the Treaty—indeed the Japanese had already complained on this score. These ships were not limited by the treaty and were proposed as a means to circumvent the fortification clause in getting air power to its most remote and needy customer—the Asiatic Fleet. Also, the navy investigated long-range, lighter-than-air craft—a pet project of Admiral Moffett's. Moffett also supported the development of long-range, land-based patrol aircraft. Moffett's accidental death in 1933 did not halt these other aviation programs.[9]

The navy's frustration with the slow introduction of operational aircraft carriers to the fleet led Moffett to propose—and the chief London naval advisor, Admiral Pratt, to support—the insertion of language into the London Naval Treaty that authorized a new hybrid class of vessel, the flying deck cruiser. The flying deck cruiser was viewed by Moffett, Pratt, and the General Board as another way to get aviation out to the fleet. Smaller ships, authorized by the Treaty, might be easier to build in the fiscally constrained political environment that had prevented the building of aviation tenders and slowed the introduction of carriers into the fleet to a crawl. Moffett also argued that flying deck cruisers might prove to be a more expedient means of generating air power for the fleet, especially its scouting component. More decks meant a higher sortie rate for aircraft and less risk if an aircraft-carrying ship was lost. Aviation would be a part of the fleet by hook or by crook—at least that was the hope of men like Moffett and the members of the General Board. There were dissenters. Adm. Hilary Jones believed that any money spent on aviation ships must go to aircraft carriers instead of a crazy "hermaphrodite" ship. Other admirals, few in number, remained skeptical of the claims of

Pratt and Moffett, but were willing to see aircraft carriers built since they felt the United States must maintain at least the treaty ratio of superiority in this class.[10]

Air Power and the Asiatic Fleet

The fortification clause of the Washington Naval Treaty immediately affected the naval aviation component of the U.S. Asiatic Fleet. The bulk of this fleet was based in the Philippines and its mission was primarily associated with the defense of the United States' overseas possession as well as protecting American interests in China—the "Open Door" policy. It was one thing to advocate that airplanes, submarines, and other vessels not limited by the Treaty be used to circumvent restrictions in the western Pacific, but implementing this solution was not so easy. The first efforts to strengthen the aviation of the Asiatic Fleet soon caused problems. The only airplane tender with this fleet was an ex-collier, the *Ajax*. Its shortcomings, given the realities of the Treaty, would soon become manifest. Six aircraft of Torpedo Squadron Twenty (VT-20) had arrived in the Philippines in crates in late 1923. The fleet Commander, Adm. Thomas Washington, specified that two of these aircraft be assembled on the wharf at Olongapo (Subic Bay, Philippines) and then stored on *Ajax*'s deck. The remaining planes and spare equipment were to be stowed below deck. The following report filed by Admiral Washington highlights the extreme problems associated with this relatively innocuous task:

> *In view of the Washington conference treaties* [sic] *rigid instructions were issued* that no aviation gear was to be kept ashore, that personnel attached to the squadron was to mess [eat] on board the tender, and that shore facilities were to be used for repair work such as is customarily handled for the ships of the fleet by the shore establishment. Olongapo was best suited to this work of erection, but the *Ajax* was withdrawn from that port as soon as the necessary work of erecting and testing the two 'planes was completed. *In spite of the care exercised to avoid any increase in the shore establishment, complaint was promptly made by the Japanese government that an aviation base was being established at Olongapo in contravention of the treaty* . . .[11] (emphasis mine).

This excerpt makes it clear that any aircraft supporting the Asiatic Fleet must be kept afloat. The only way to do this without an aircraft carrier was to use tenders. As with submarines and destroyers, this class of naval auxiliary included spare parts, machine rooms, and berthing for additional crew or replacements. Airplane tenders also needed cranes to hoist the planes on and off the ship.

Airplane tenders had been a component of the building program since the first *U.S. Naval Policy* had been approved in 1922. Both OpNav's War Plans Division and BuAer recommended that tenders be built or converted from existing ships. War Plans recommended seven for wartime use and Admiral Moffett revised this figure down to a peacetime building or conversion program for five "air depot ships." Initially the justification had not been specifically for the purpose of forward-deploying aviation to the Asiatic Fleet, but rather as another way to support the fleet. Battleships and cruisers had very little room aboard for aircraft repair and parts storage. Lacking maintenance facilities ashore in the distant waters of the Western Pacific, airplane tenders could provide the fleet the logistics needed to support naval aviation afloat. Accordingly, there were two roles for these ships: first, to service those planes embarked on the ships of the battle and scouting fleets and secondly, to serve as "floating air bases" for the aircraft of the forward-deployed Asiatic Fleet. The legality of building these ships under the terms of the Treaty came up during a March 1922 General Board hearing. Admiral Pratt noted: "The depot ship won't be touched by the Treaty, because it comes in under the class of ship, namely: auxiliary, which isn't touched by the Treaty . . . you can have them as large as you choose and as many as you choose."

However, when Captain Henry C. Mustin of BuAer added that airplane tenders should also carry "flying-off decks." Admirals Pratt and Rodgers suggested that this was in fact a violation of the Treaty. Subsequently the General Board proposed that five tenders be included in the building policy section of the 1922 *Naval Policy*. This figure was reduced to two tenders "for aircraft" due to worries by Navy Secretary Denby that Congress would balk at the number of ship types not "touched by the Treaty" being recommended by the Board.[12]

BuAer added its voice to that of the Asiatic Fleet commander (CINCAF) in complaining about the limitations of the *Ajax*. In a June memorandum prepared for the CNO, BuAer noted that *Ajax* hosted five DT-2 aircraft (seaplanes); of these only two were operational. The memo flatly stated that "The U.S.S. *Ajax* . . . is entirely inadequate." It then went on to note that the total aircraft force of the Asiatic Fleet would eventually number 72 aircraft divided into four squadrons, but that *Ajax* was not prepared to handle the eighteen assigned to her. All these facts were subsequently aired before the General Board. CINCAF requested that the Board discuss the proposal to assign two airplane tenders to his fleet. The Board used this opportunity for "the consideration of a new type of ship for the Navy and that is the aircraft tender."[13]

No tenders had yet been built from the keel up, although two ships had been converted with various degrees of success: the relatively suitable *Wright*, which could handle 24 planes, and the abysmal *Ajax*. *Wright*, however, was not assigned to the Asiatic Fleet. BuAer had suggested a general design goal that would result in a ship that could host 36 aircraft (two squadrons) and the General Board had duly recommended that two such ships be acquired. BuAer raised additional issues in October 1924, referring again to *Ajax* as "entirely unsuited" to its function. Interestingly, the tender was one of three options. BuAer thought a naval air station most suitable for the Asiatic Fleet followed by a carrier. Ideally this fleet would have both, however, the October memo wryly noted that "After a study of Article 19 . . . of the Limitation of Armaments Treaty . . . it is not believed that there is anything which prevents the use of . . .facilities [at Paranque in the Philippines] by VT-20 and it is recommended that the use of these facilities be taken up with the Philippine government by the commander-in-chief, Asiatic." Nothing came of this initiative but the Board suggested to the secretary that the Treaty did allow the "repair and replacement of worn out weapons, and equipment," such as aircraft, as long as "existing naval facilities" ashore were not increased. Given Japanese scrutiny of the Philippines, it was unlikely they would have allowed this solution to be implemented without protest.[14]

The Board could either convert additional ships to airplane tenders or it could recommend the design and construction of an entirely new ship. On December 10, 1941, the General Board held a hearing on the tender issue. Admiral Beuret of BuC&R recommended an entirely new design. Admiral Strauss, who facilitated the discussion for that day, tended to favor conversion. The following interchange from that hearing is quite revealing:

> Commander Whiting: With a tender you could move from place to place, whereas *if you don't have tenders you have got to build bases.*
>
> Admiral Strauss: *This is a mobile station.* I understand the basic question in Admiral Beuret's mind is this tender to transport the airplanes to a long radius, then put them in the water, or is it to tend the airplanes that fly from point to point and look out for them when they get out to the new points?
>
> Admiral Beuret: Yes, that's it.
>
> Commander Whiting: Its for the purpose of *operating* with the fleet where they are *operating*, but not to accompany them on the high seas.[15] (emphasis mine)

These officers realized that they must account for operations without bases in the design of their ships for the Far East. More and more the solution had to be a "blue water" concept. Logistics support for the fleet would have to come from what was called the fleet train—the auxiliary oilers, tenders, ammunition ships, and so on—

and through the use of mobile bases, in this case airplane tenders. The emphasis on the operational nature of the solution is reflected in Whiting's comments. The tenders were not to accompany the fleet into battle but to set up within the theater of operations to support the squadrons. These squadrons had the tactical role of supporting the fleet in battle as well as providing reconnaissance and defense for the fleet train.

The outcome of this hearing was, typically, a compromise—since both short- and long-term problems needed to be resolved. Because conversion of an existing ship was cheaper, it was decided to replace *Ajax* by converting the collier *Jason*, a ship similar in size to the aircraft carrier *Langley*, into a tender for the Asiatic Fleet. Additional ship conversions were also planned. Finally, in order to have tenders for the remainder of the fleet and solve the long-term problem of how to accommodate 72 planes in peace and 126 in war, the General Board recommended construction of a purpose-built airplane tender. As fiscal realities came into play the only thing accomplished was the dispatch of the *Jason* to the Asiatic Fleet.[16] It was by this means that CINCAF received another airplane tender. *Ajax* was relieved of all duties and decommissioned upon *Jason*'s arrival at Cavite in late May 1925. *Jason* served as an airplane tender for the Asiatic Fleet without receiving any of the modifications recommended by the General Board, probably due to her busy operational schedule supporting the Marines in Nicaragua and sparse funds for naval reconstruction. These funds were needed to complete the conversion of the two battle cruisers into aircraft carriers.[17]

By 1928 the naval air situation in the Asiatic Fleet had become dire. The Commander of Aircraft Squadrons, Asiatic Fleet, Cdr. R. K. Turner (later of World War II fame), forwarded his concerns to fleet commander Adm. Mark Bristol in March 1928. Turner was alarmed about the maintenance of fleet aircraft in general and the suitability of *Jason* to support his planes in particular.[18] Turner noted that the "Treaty does not allow an increase of naval shore equipment in the Philippines." He proposed that "a well equipped aircraft repair base in the Philippines, available for both Army and Navy planes" be designated to overhaul his seaplanes. The humidity of the tropics, plus the necessity of leaving the seaplanes in the salt water anchorages due to the treaty, had drastically shortened the time between major overhauls for these aircraft.[19] Even so, *Jason* could barely handle six planes in her current configuration and without support ashore. Turner closed his communication by recommending that CINCAF contact the Army Governor General of the Philippines to arrange the use of their aircraft maintenance facilities for the planes. Simultaneously Turner contacted local Army officers on his own initiative. Bristol noted as much in an April endorsement of Turner's report to Admiral Moffett at BuAer: "arrangements have been completed with the Army for the overhaul of Navy planes at the Army base."[20]

Bristol and Turner had jumped the gun with the Army. By August their correspondence to BuAer reflected that the idea of using Army facilities to perform depot-level maintenance on the seaplanes had not been implemented. Turner informed BuAer and CINCAF that his planes were deteriorating rapidly due to a lack of space aboard the *Jason*. The planes were "at the end of their useful life" and as a result "of little use as heavy bombers or torpedo planes." Turner asked if Moffett could initiate communications with the Army department to help get the Asiatic fleet planes overhauled. Otherwise the planes would have to be shipped back to Pearl Harbor with no replacements. Turner proposed a number of modification plans for *Jason* in addition to his request to use Army facilities: "No more than six VT planes can be tended by the JASON before alterations are made. Partial alterations will allow twelve planes to be tended. Complete alterations will allow eighteen VT's [torpedo], six VO's [observation], and six VF's [fighter] to be tended. VP [long-range patrol] planes should be based on shore, but lacking facilities, are unsuitable for the Asiatic."[21]

Admiral Bristol dutifully endorsed Turner's recommendations to BuAer, both for use of the Army facilities and the comprehensive modification of *Jason*.[22]

The effort in the Navy to solve this problem moved to the next organizational level—the General Board, BuAer, and OpNav. By November Moffett forwarded all the relevant correspondence to the CNO (Adm. C. F. Hughes) who forwarded much of it to the General Board. The Board was hard at work calculating budget allocations for the final form of the Butler "cruiser bill." Perhaps some of this money could be diverted to convert *Jason* into a more suitable airplane tender. Moffett added his assessments and recommendations to the mix. He stressed that the correspondence dealt "with the questions of number and types of airplanes necessary to form a minimum *balanced* air force for the Asiatic Fleet and the basing facilities necessary to support them." (emphasis mine) Moffett knew the word "balanced" would resonate with his audience. Admirals Hughes and Jones, both members of the General Board, were Treaty-savvy, surface admirals who saw aviation as a broad solution, not just limited to use on aircraft carriers but integral to the entire spectrum of fleet operations. Moffett stressed that "Present treaty limitation [the fortification clause] . . . in effect, require[s] that such bases be self-contained floating units." He also highlighted the impact of the situation on an area directly under the CNO's purview—war plans. The BuAer chief pointed out that the final total of airplanes required to support current war plans was thirty-six, far more than *Jason* could support under present conditions. Moffett cited the "Revised Basic Orange Plan" which would "require that these [floating] bases possess a considerable degree of mobility." This sort of mobility required ships. For good measure he threw cold water on the War Plans Division's proposal to also base the larger long-range patrol (VP) seaplanes with the Asiatic Fleet, saying they were unsupportable without shore basing.[23]

Admiral Moffett did not agree with Turner's proposal to use Army facilities since it required Army money and "goodwill" that could be removed at any time without the Navy's concurrence. His preference was to do what the British had just done with their old experimental carrier *Hermes*—base an aircraft carrier in the Far East. Perhaps the experimental *Langley* could be used as a "floating air base" for the fleet's planes. *Langley* had sufficient deck space for in-theater maintenance and it also had the advantage of providing a ready flight deck for use as a conventional carrier in war. According to Moffett, carriers were "the proper and most effective bases" for the Asiatic Fleet. "In the absence" of carriers, new airplane tenders seemed "the most suitable alternative." The last option was to modify *Jason* to support more planes. Moffett scaled back the recommendation of CINCAF based on "the limited number of active service years remaining" for *Jason*. *Jason* would be modified so that she could support twelve torpedo planes. In addition, Moffett recommended that the Navy "Convert an oil tanker of suitable type for aircraft carrier service, and install major repair facilities on it." The nine-page message closed ominously: "To this Bureau it does not appear that there can be any satisfactory solution under present restrictions, to the problem of establishing an effective Aircraft squadron in the Asiatic that does not include the assignment of a carrier to that station."[24]

A partial solution to the immediate problem was implemented in 1929. Money was allocated to convert *Jason* in the Philippines to allow her to properly tend six torpedo planes and six smaller observation planes, not the twelve torpedo planes considered the absolute minimum. *Jason* was formally designated as an airplane tender (AV-2) in 1930.[25] The tender option remained problematic for the remainder of the Treaty period. The Navy refused to spend its limited budget on tenders while it still did not have its proper ratio of treaty-allowed aircraft carriers. By 1933, the General Board, now headed by the same Admiral Bristol who had been CINCAF, recommended against building any "fleet auxiliaries [including airplane tenders] until the Navy has been built to Treaty strength in the Treaty limited classes."[26]

Moffett never lived to see it, but portions of his preferred plan were executed prior to World War II. *Langley* was designated a tender (AV-3) in 1937 and eventually assigned to the Asiatic Fleet (1939) in precisely the role he envisioned. Additionally, purpose-built airplane tenders were finally constructed for the seaplane fleet, to include a capability to support the larger VP seaplanes, but only after the treaty system had effectively lapsed. However, instead of being deployed to the Philippines, the tenders were retained as mobile bases to accompany the main fleet, which would be used in the more deliberate advances in the South and Central Pacific. These tenders eventually provided a critical basing capability for forward deployed seaplanes like the PBY Catalina that was critical during the conduct of the early campaigns of World War II, especially at Guadalcanal.[27]

The case of the airplane tenders for the Asiatic Fleet highlights the extreme urgency within the Navy regarding the need for carriers and naval aviation. Also, the proposal to create a new class of ship not limited by the Treaty illuminates the role the Treaty played in channeling Navy leaders, sometimes against their inclination, toward nontraditional solutions to their problems. Moffett certainly felt that a carrier was the best course of action for the Asiatic Fleet but was willing to compromise in order to get air logistics support afloat. Admiral Bristol, on the other hand, clearly favored leveraging the support of the Army and an extensive program of modifications for *Jason*. By 1933, Bristol, who had reported for duty to the General Board first in 1930, had come around to Moffett's way of thinking. After joining the Board, Bristol would have been subject to Moffett's constant advocacy, both written and oral, for carrier aviation. On the other hand, Adm. Ernest J. King, Moffett's successor at BuAer proposed building more aircraft tenders *and* building up to carrier limits.[28]

What is surprising is how determined, optimistic even, some Navy leaders remained about getting these ships built or converted in the austere fiscal environment of the Hoover administration. The search for government economies even caused Hoover's budget director H. M. Lord to forward a congressional proposal by Representative French to lay up the newly commissioned large aircraft carriers to save money on their "stupendous" operating costs. Another memorandum from Hoover to his Navy Secretary C. F. Adams points out a dilemma in the way program funding, the Depression, and the treaty system intersected: "it does not seem to me feasible to undertake the reconstruction of the three battleships at the present time if we are at the same time to undertake work on new treaty construction. We are faced with a deficit of 400,000,000 dollars in the government, and I am compelled to choose such items as will give the largest measure of immediate employment."[29]

Proposals and decisions such as this tended to cause officers like Bristol and Moffett to rally around the most important programs in an effort to avoid losing more ground that advocacy of a wider range of programs might have caused. However, it is a testament to the interwar Navy that in spite of these difficulties they often recommended year in and year out the building of interesting and unique *logistics* ships like airplane tenders. By 1940, the aviation tender blueprints, long in mothballs, had been used to build new vessels like the seaplane tender *Curtiss*.

The Effect of the Treaty on Aircraft Carrier Introduction

Some have criticized the interwar Navy for its slow pace in introducing aircraft carriers into the treaty fleet. They say this is evidence that the interwar Navy was lackluster in its enthusiasm for a revolutionary new concept. However, parsimony and the supposed lack of interest in aircraft carriers were not the only factors that delayed the building them for the treaty fleet. The Washington Naval Treaty also played a role in slowing the delivery of carriers to the fleet by interacting with the tendency of the interwar administrations (until FDR) to take advantage of opportunities to economize while at the same time appear sensitive to the spirit of the treaty system. The United States could save money while providing a "good example" by restraining its naval construction, especially the "new" aircraft carriers *Saratoga* and *Lexington*.

The Navy's difficulties in converting the *Saratoga* and *Lexington* into aircraft carriers were partly self-inflicted. These difficulties had their origin in the Navy's liberal interpretation of the Treaty's articles on aircraft carrier construction and the reconstruction clause. Article IX of the Washington Treaty allowed the United States to "build not more than two aircraft carriers, each of a tonnage of not more than 33,000 tons . . . and in order to effect economy any of the Contracting Powers may use for this purpose any two of their ships . . . which would otherwise be scrapped under the provision of Article II." The Navy added 3,000 tons to the 33,000 tons for "bulge or blister and anti-air deck protection" as specified in the reconstruction clause. The extra 3,000 tons was critical because the Navy had already decided to convert several of its battle cruisers to aircraft carriers to save costs prior to the Washington Conference. These plans had envisioned a ship of approximately 39,000 tons displacement but could be scaled back to 36,000 tons according to BuC&R.[30]

The General Board began consideration of these design proposals in early February 1922—prior to congressional ratification of the Treaty. The only Treaty consideration that arose was *how to* use the 3,000 tons, not the legality of doing so in the first place. Admiral Pratt emphasized that this additional tonnage must be used "in blister or anti-air attack deck protection or you can't use one pound of it." Accordingly, the Navy proceeded "full-steam ahead" with 36,000 tons as its tonnage cap for the new ships. Because the reconstruction clause applied to capital ships, which *Saratoga* and *Lexington* nominally were, the Navy only counted them as 33,000 tons apiece against the 135,000 tons allowed it by the Treaty. This view originated in the General Board, which claimed that the additional 3,000 tons per ship were allowed by the Treaty but did not count against the overall carrier tonnage![31]

The Board took the additional precaution of soliciting views from one of the legal advisers to the U.S. delegation at the Washington Conference, Prof. George Wilson of Harvard University. Wilson confirmed that the negotiations about the language of the reconstruction clause in January 1922 during the Conference had been "understood" to authorize "an increase to the allowed tonnage, but that such increase should not exceed three thousand standard tons (3,048 metric tons) for each ship."[32] The Board released its final judgment on the issue to the Navy secretary summarizing that their position was that the 3,000 additional tons per ship could be built without deducting them from the 69,000 tons remaining for carrier construction. The Navy could have its cake and eat it, too—courtesy of this innovative interpretation from the General Board.

All seemed well through 1924 as conversion of the two battle cruisers proceeded apace. The only problem that was experienced was the same one experienced by all the naval construction programs—the traditionally slim support for military programs by the Republicans then in power. Funding, and inevitable cost overruns (often due to limited, inefficient funding), slowed the pace on the completion of these ships.[33] In late 1924 the General Board emphasized the Navy's position that the two 35,000-ton plus carriers accounted for only 66,000 tons against the Treaty limit of 135,000 tons. At the same time the clamor for these ships to join the fleet increased.[34] Representative Burton French, Chairman of the House subcommittee on Naval Affairs, Admiral Moffett, CINCAF, CINCUS, the War Plans Division, and the General Board all demanded authorizations for additional carrier construction to begin immediately. Every annual General Board Building Policy until 1928 (the year *Saratoga* and *Lexington* officially became operational) emphasized the urgent need to complete the construction of the two large carriers plus additional allocations for new carriers. French was "disturbed" over reports (from BuAer and the General Board) that the Navy "had retrogressed" vis-à-vis the carrier forces of Japan and Great Britain. Also, both the Board and BuAer pointed out that the slow pace of conversion of the battle cruisers had delayed developing experience with the large carriers needed as feedback for the design of any new carriers. The Board, noting the delay in fielding the larger ships, argued that the Navy should not delay any more the building of a smaller carrier because it perceived the U.S. position to be inferior to that of Great Britain and Japan. The Board also pointed out to the Navy secretary that additional carriers were needed "as a means of transporting aircraft" because "we are debarred from establishing naval airplane bases in the Philippines."[35] This last function absolutely predicted the vital role Pacific Fleet carriers initially played after Pearl Harbor in building up the air forces shattered at Pearl Harbor and in the Philippines. These arguments did not persuade President Coolidge. His proposed appropriations for the 1926 budget included funds only for the "continuation of the construction of two aircraft carriers [*Saratoga* and *Lexington*]."[36]

It was at this low point, in 1925, with the pressure to build more carriers contingent on the behind-schedule *Saratoga* and *Lexington* that the Treaty came again into the foreground. The protest by Great Britain over turret modifications for battleships became conflated with the size of aircraft carriers in a most unexpected way. Coolidge's new Secretary of the Navy was Curtis Wilbur, a man who received his appointment as a result of the Teapot Dome scandal and who was charged to help keep the Navy "above board."[37] The legalistic secretary, in his zeal to maintain the spirit of the Treaty and the reputation of the Navy, not only prevented the funding of the requested turret modifications but also asked that the Board "take up the question as to what changes in these ships [*Saratoga* and *Lexington*] should be made in order to reduce their tonnage to 33,000 tons." Wilbur also directed that all construction for emplacing guns and additional armor on these vessels cease until he got his answers. Wilbur took the exceptional action of directing the General Board to forward all dissenting views with the majority report.[38]

The General Board immediately convened hearings to address Wilbur's guidance. After the hearings, the Board notified Wilbur that significant power reductions would be necessary as well as the removal of some of the ships' side armor to get their weight down to 33,000 tons. The Board also restated the source of their legal opinion. Three thousand tons was completely in accordance with the Treaty and aligned with the recollections of key Washington Conference delegates and advisors: Professor Wilson, former acting secretary Theodore Roosevelt Jr., Admiral Taylor, and Admiral Pratt.[39] Wilbur recalled Pratt, who was serving as the commander of Battleship Division Four with the fleet, and hurriedly assigned him for service with the General Board again. Pratt's real mission was to visit former Secretary of State Hughes in order to get his opinion on the legality of adding the extra three thousand tons. Pratt was also to add his own assessment to that of Hughes and all of this was to be provided to the General Board and then to the secretary. Evidently Wilbur was hoping either Hughes or Pratt would dissent from the General Board's position—or perhaps cause the General Board to change its position based on the authoritative opinion of Hughes. Pratt and Hughes met on June 24, 1925 at Hughes' retreat in upstate New York. They agreed with the Board that the three thousand additional tons were authorized and could be "liberally" counted against weight increases due to internal compartmentalization changes for fuel storage. These changes could be characterized as anti-torpedo defense measures. Wilbur bowed in the face of the logic of the illustrious statesman, although the additional cost of undoing the construction already accomplished on the two carriers was also a factor. The net effect of Wilbur's intransigence was to delay still further the operational introduction of *Saratoga* and *Lexington*.[40]

The Flying Deck Cruiser

The slow pace of introducing aircraft carriers was not the only problem the Navy faced as it approached the 1930 London Naval Conference. Different ways to ameliorate the fortification clause had proved inadequate as far as the General Board was concerned. Airplane tenders, shared service maintenance, or the assignment of an aircraft carrier to the Asiatic Fleet—were all partial solutions and workarounds. Even the battle fleet was deficient in its aviation support due to the slow introduction of carriers. The stubborn commitment of the administrations of the 1920s to reduced naval construction in the hopes of gaining additional agreements for naval limitation, and thus reduced construction requirements, had seriously compromised the Pacific strategy reflected in Orange War Plan. There were those in the Navy who felt that even the defense of Hawaii and the Panama Canal were now in jeopardy. The 1930-era Orange Plan relied on maximum naval construction for treaty-limited aircraft carriers and liberal construction for nonlimited naval construction like submarines, cruisers, auxiliaries (e.g., tenders), and ambitious afloat logistics programs like the mobile base project (see Chapter 7). The Treaty also did not limit the numbers of naval aircraft that could be built.[41]

Plans and blueprints existed on paper, but without actual platforms with which to experiment, the Navy felt it had no solid basis of evidence on which to test its plans as to their feasibility or suitability. Neither could their associated technology and ship designs be evaluated in the maritime environment. Increasingly, Navy leaders tried budget strategies that would get platforms built—any suitable platform—so they could at least proceed with realistic experimentation using fleet battle problems and war gaming at the Naval War College. Navy leaders also continued to spread solutions across the force. The uncertainty about the character of a future naval war involving air power with almost no land bases, combined with the late introduction of the *Saratoga* and *Lexington* to the fleet, had undermined these leaders' confidence about how to structure and base naval air with the fleet. It must be afloat due to the fortification clause, but on which ships and in what numbers? Lack of experience with *Saratoga* and *Lexington* had led BuAer, in particular, to start to question the advisability of putting the majority of the airplanes on a few large carriers. Additionally, the wisdom of putting the balance of these ships in close proximity or direct support of the battle fleet was also being questioned, and not just by Moffett but by other admirals like Pratt, Hilary Jones, J. M. Reeves (first captain of *Langley*), Frank Schofield (head of the War Plans Division), and Mark Bristol (CINCAF).

It was against this background that the idea for a flying deck cruiser was inserted into the language of the 1930 London Naval Treaty. Such a ship would serve multiple purposes. First and foremost it would get naval aviation to the fleet by employing a different class of ship—ships that could carry more than just two or three planes but not nearly the numbers of the two large carriers. Such ships promised operational flexibility because aviation would be distributed more evenly throughout the fleet. The risk of losing one of the two big carriers as had happened in the famous Fleet Problem IX, and thus losing a significant percentage of the aviation available to protect and assist the battle line, would be ameliorated by the addition of such ships. Spreading aviation throughout the fleet seemed to make sense given the limited evidence available at the time.[42]

When viewed in this light the flying deck cruiser was a conceptual breakthrough rather than a mere tactical improvement. The Navy approached the design for these ships with an open mind. The General Board, as usual, controlled the process considering designs ranging from standard cruisers with just a few additional aircraft to flush deck designs almost indistinguishable from the sorts of light aircraft carriers that were later developed from cruiser hulls for World War II. Cruisers had always been the more expendable units, meaning suitable for riskier missions, in the fleet when compared with battleships. However, aircraft carriers were clearly not expendable in the way cruisers were. Because of its limited numbers, and the threat land-based aviation posed to the fleet, the carrier was considered by many officers to be a high value platform, one that the Navy could not do without and still win. The flying deck cruiser, on the other hand, might be lost and the fleet could still function—especially if large numbers of them were built. The construction of a number of them might even increase the survivability of the more precious carriers.

Because the cruiser was smaller and cheaper per unit, it was hoped more of them could be built than had been the case with aircraft carriers. More cruisers meant that they could be assigned to provide, or dispute, command of the air in the remote areas of the Pacific and provide cover for those assets for which larger carriers were considered inappropriate (such as the fleet train or portions of the scouting forces). As a cruiser, the ship could also perform all those functions—screening, reconnaissance, and commerce raiding—at which cruisers excelled. It could also do these missions and new ones like convoy protection and antisubmarine warfare, in ways that aviation could only enhance because of its operational reach. Instead of defending or screening within line-of-sight of a convoy or patrol area, the cruiser could extend its search area over hundreds of square miles. Using its complement of aircraft, it could also strike at long ranges—a very useful function if a lone enemy submarine or ship was located. Finally, it would be a "Treaty" ship, one that naval officers could point to in the language of the London Treaty in order to justify construction. It stood a good chance of being constructed by stingy administrations and Congress because it was relatively inexpensive. As it turned out, Congress authorized the construction of the flying deck cruiser before the Navy was ready to build it.[43]

Admiral Moffett had grown frantic on the issue of carrier introduction by the late 1920s. He had asked the Navy secretary to recommend that five carriers of the smaller size be built and had only gotten one. The secretary duly forwarded Moffett's correspondence to the General Board. Moffett's change of heart on the size of carriers had to do with his increasing conviction that numerous small carrier decks might in fact be better than large decks in order to generate more sorties (flights). Moffett felt the situation was so dire that he began to seriously investigate using large dirigibles as a kind of airborne aircraft carrier.[44] He was also coming to believe that spreading aviation around on smaller decks would allow the Navy more operational flexibility in using these ships. The General Board had concurred with most of this reasoning, although it was loath to abandon the big carriers until more evidence was available after experience was gained with the smaller *Ranger* in the fleet.[45] With only the old, slow *Langley* there were hardly enough carriers to support the battleships. Even with every ton built, Admiral Pratt—while president of the War College—became convinced that there were not enough carriers to do everything that needed to be done. According to a 1927 annual fleet report: "in a Pacific War, Navy control of the air cannot be attained by any arrangement of tonnage of aircraft carriers within the allotment of 135,000 tons, whether all were of 10,000 tons, or all of 27,000 tons displacement, and that in order to secure this, quantity production will have to be resorted to."[46]

Reports like these were slowly convincing Moffett, and some members of the General Board (of which Pratt was a member), that some mechanism for introducing additional aviation into the fleet—lacking bases and carriers—must be achieved. Admirals Bristol and Moffett had emphasized the urgent situation of the Asiatic Fleet and its bearing on strategy and were supported in their views by Pratt: "One of the outstanding lessons of the overseas problem played each year is that to advance into a hostile zone the fleet must carry with it an air force that will ensure, *beyond a doubt,* command of the air. This means not only superiority to enemy fleet aircraft but also to his fleet and shore based aircraft combined."[47] (emphasis original)

Admiral Schofield, now chief of the War Plans Division, added his voice to the clamor for more carriers. In a memorandum addressing the aircraft building program, Schofield used the occasion to bluntly state that "I consider the greatest need of naval aviation today is more carriers."[48]

Moffett's memorandum of July 1928 received very broad dissemination. He included a complete history of what he regarded as budgetary neglect for aviation shipbuilding. Moffett cited almost every significant organization in the Navy, to include the Naval War College, CNO and the War Plans Division, the General Board, and the commanders of the U.S. Fleet and his subordinates in the Battle, Scouting, and Asiatic Fleets, to show that his ideas had broad support throughout the Navy. The memorandum also showed that Moffett was not alone in identifying the critical need for naval aviation as an operational means to enable the strategy of the Orange Plan by ameliorating the prohibition against basing aviation ashore in the western Pacific. Buried in Moffett's memo was the following: "Within ten thousand tons displacement it is practicable to build an aircraft carrying cruiser which can operate upward of forty planes of the intermediate or smaller types. Such a complement of planes can be used alternately on gunnery observation, tactical scouting, fighting [air defense], smoke laying or bombing missions with bombs carrying at least four hundred pounds of explosive." Here was the outline for a new type of naval vessel—a hybrid cruiser/aircraft carrier.[49]

Moffett went on to outline the operational impact such a vessel might have by positing what might happen to a U.S. " . . . 8[-inch] gun cruiser" force that encountered "enemy cruiser forces composed of a combination of 8[-inch] gun carriers *and* light aircraft carriers." (emphasis mine) Moffett framed the operational uses of such a force by asking "what tactical methods" would offset a combination of cruisers and light carriers:

1. To offset the decided advantage in the service of information and security, which goes with superiority in air strength?
2. To offset the decided advantage in effective striking range inherent in large numbers of bombing aircraft?
3. To offset the decided advantage in flexibility of operations which is inherent in the high speed, relatively long striking range and maneuverability of accompanying air forces?
4. To offset the decided advantage in protection, which is inherent in the use of smoke laying aircraft for concealment and observation?[50]

Moffett was clearly worried that the Japanese might take this path given their ambitious cruiser construction programs. The Japanese, in fact, did lay down scouting cruisers with an increased air complement beginning in 1934—the *Tone* class.[51]

Moffett's solution was a new vessel with the 10,000-ton displacement of a cruiser that carried substantial numbers of aircraft. If one could not have light carriers to accompany the cruisers, simply turn the cruisers into light carriers in all but name. The flying deck cruiser might be legally acceptable under the terms of the Treaty because it had been vague on just what defined an aircraft carrier. Moffett's arguments were meant to highlight, under the current treaty system, that the Navy had to go beyond building just its allotment of carriers. It had to develop a completely air-integrated fleet. The frustrations of the 1920s in trying to employ naval aviation as a part of a system of solutions to address the lack of naval bases in the Far East, ashore and afloat, had led to a point where other naval leaders were open to ideas that might otherwise have been rejected out of hand—either as wild, hybrid schemes, or as violations of the Treaty.

The idea of putting more than just a couple of spotting or reconnaissance planes on cruisers, the most significant class of ships whose quantity was not limited at Washington, had been around for a while. The Navy had always intended, after the first experiments, to put airplanes on cruisers—specifically sea planes. These would fulfill a variety of roles, chief among them spotting and reconnaissance due to their limitation as high performance fighters, although this role was considered, as well. During the early hearings on the construction of aircraft carriers in 1923, Admiral Rodgers had emphasized the loophole in the treaty that allowed the Navy to "make as many ten thousand ton carriers as we like. That is to say ten thousand tons without fuel, the same as we can build cruisers. . . . " Note how Rodgers hinted at stretching the allowable displacement by not including fuel. Rodgers was supported in this view by Admiral Strauss, the next senior member of the General Board. Admiral Moffett, also at this hearing, stated that he "would be in favor of making the larger ships owing to the fact that the larger ship would have a steadier platform for landing." Moffett, as we have seen, changed his views substantially as the fleet gained more experience with deck landings and as technological solutions were implemented. However, based on Moffett's 1923 testimony, the Board had shelved the idea of 10,000-ton "airplane carriers" in favor of larger 23,000 to 27,000-ton designs. Moffett, however, did not forget Rodgers' 10,000-ton proposal as it applied to cruisers.[52]

The General Board held hearings to examine various design considerations for 10,000-ton cruisers in 1925. At these hearings Capt. William Standley of the War Plans Division presented a study on the use of light cruisers in a future war based on the experiences of World War I and the requirements of the Orange Plan. Standley emphasized that any ship was a "complex of compromises" and then listed the characteristics that War Plans wanted the cruisers of the fleet to have in order to execute the war plan. One of the characteristics listed was the embarkation of "combat [fighters], scouting and bombing airplanes." Standley later qualified this characteristic as being "provision for as many airplanes as possible consisting of both combat planes and combined bombing and scouting planes." In addition, these cruisers should have "a flying deck for launching and receiving airplanes." The Board's response to this proposal was tepid. They were still unsure as to whether the Treaty would allow such a thing.[53]

Prior to the London Naval Conference of 1930, Moffett lobbied the General Board to recommend formally that the London delegation propose "transferring [*Lexington* and *Saratoga*] to the experimental class" of carriers allowed by the Washington Treaty. In this way the United States could replace these two carriers with four "16,500-ton carriers" that had "several advantages over" the larger carriers. Moffett's intent was to try to retain the large ships while building even more of the smaller ones. The General Board forwarded Moffett's advice and endorsed it by recommending that should the naval conference in London advocate a reduction in overall carrier tonnage "below 135,000, the General Board is of the opinion that the United States should insist that the *Saratoga* and *Lexington* be placed in the experimental class as a necessary provision for agreement." The bottom line was that treaty allowances for carrier tonnage should not be decreased under any circumstances since "The United States must conduct any naval war at greater distance from home than would be the case of other Powers signatory to the Washington Treaty."[54]

The London Conference results, in part because of the recommendations of Moffett and the General Board, did not reduce the overall carrier tonnage or ratios codified at Washington—the U.S. allowance for carriers remained 135,000 tons. However, the London Treaty specifically defined aircraft carriers in such a way as to remove the "loophole" that might allow the building of carriers of 10,000 tons or less. However, through Moffett's efforts—perhaps even subterfuge meant to be appear as "a spur of the moment" idea during the conference—a new "loophole" was created using two separate articles. Article 3 of the London Treaty prevented the counting of "landing-on or flying-off platforms" against a nation's overall carrier tonnage. Also, Article 16 of Part III of the treaty stipulated that "Not more than twenty-five percent of the allowed total tonnage in the cruiser category may be fitted with a landing-on platform or deck for aircraft."[55]

The Navy's—and Moffett's—purpose in creating the Treaty language for the flying deck cruiser has been misunderstood. Some have characterized the flying deck cruiser, along with Moffett's advocacy of lighter-than-air dirigibles, as examples where he was "wrong." Others see the flying deck cruiser as a case where the General Board served admirably in the role as "umpire," preventing the Navy from making an "unfortunate" decision. Still others regard the concept as visionary and "before its time." They claim the flying deck cruiser might have measurably improved the performance of the Navy during the first year of World War II.[56]

What is clear is that the situation as perceived by Navy leaders in 1930 was urgent. The Navy had only just begun to experiment with the big carriers *Saratoga* and *Lexington*. The General Board was still undecided as to the utility of large carriers versus small carriers. The one thing everyone in the Navy did agree on was that there were not enough aviation ships in the fleet—carriers, tenders, or flying deck cruisers. Moffett's diary entry makes clear that the platform was secondary to his primary consideration—air power for the fleet: "Aircraft will settle the next war. Don't care how many surface vessels we build. The more the other nations build, the less they will have for aircraft. Friday Mar 21st proposed to Captain VanKeuren he get up a design for a 6-inch cruiser that would have a landing deck for aircraft."[57]

Moffett reflected the consensus in the Navy regarding naval aviation. Whatever means that could be used should be used to get aviation ships. Any attempt to do so must account for the constraints of the naval treaties, support by the administration, and support in Congress. Although Moffett could no longer purposefully build small carriers because of the London Treaty, he could build flying deck ships as long as they looked like, and were armed as, 6-inch cruisers—they would have to be half carrier and half cruiser. Moffett's view carried with it certain risks. First and foremost was the risk that in building flying deck ships other aviation ships might not get built. Clearly, Moffett wanted aircraft carriers as well. The concept, however promising, remained unproven until one of these ships could be built and tested with the fleet. It was only by the "slimmest of circumstances" that the prototypes for the flying deck cruiser were not built.[58]

Work began almost immediately on this concept. Moffett forwarded a memorandum to the General Board urging "that the United States proceed at once with the building of all tonnage that can carry substantial numbers of aircraft . . . that will permit its completion by the end of 1936." Included in Moffett's memo was a plan to build seven flying deck cruisers. The Board had first learned of the Moffett's proposal during an April 1930 hearing on conventional cruiser design. Moffett's representative at this meeting was Cdr. Richmond K. Turner, the same officer frustrated in the Asiatic Fleet with the aircraft tender situation. Turner read into the record Moffett's memo stating that "aviation gives every promise of extending its usefulness and of becoming a more and more effective *offensive* and defensive weapon. . . . " (emphasis mine) Turner read that the "probable future ability of planes based on ten thousand–ton cruisers to reduce the efficiency of enemy light forces by bombing is great. . . . " When Moffett forwarded his ideas directly to the secretary of the Navy, the General Board commented in its serial that "Such a type [flying deck cruiser] has not yet been considered by the General Board as no plans for such a vessel have as yet been developed." The General Board was committed to the conventional aircraft carrier and strongly recommended approval only of Moffett's proposal to build five 13,800-ton carriers by 1936. Nevertheless, the Board did not dismiss the flying deck cruiser entirely: "A vessel of such a type is not at this time believed to best meet service needs; final opinion must be reserved until developed plans are ready and further study of the subject is made."[59]

On September 17, 1930 William V. Pratt, who had been the lead technical advisor to the London Conference delegation and was currently serving as CINCUS, became chief of naval operations. His enthusiasm for the flying deck concept was nearly as great as Moffett's. Pratt had let it be known in November 1930 that "there is need for the development of this type." Additionally, he criticized some admirals for their use of the word "hybrid" in referring to the concept because it "could in fact be used to discredit any of the combatant types now in existence." These views were forwarded to the General Board.[60]

Adm. Harris Laning, the new president of the Naval War College, visited the Board on November 5, 1930 to discuss "naval subjects and the general subject of coordination and work of the General Board and the Naval War College." The flying deck cruiser was probably discussed. Shortly after, Adm. Mark Bristol, the chairman of the executive committee of the Board, referred the Navy secretary to legislation that authorized the construction of six light cruisers. Bristol reported that "If a satisfactory design of a cruiser with landing-on deck is produced the General Board anticipates recommending the construction of one such cruiser in order that its value to the fleet may be thoroughly tried out and proved." The Navy was in the unusual position of having the money already authorized to build a ship for which it had no design.[61]

Admiral Pratt made his views on the flying deck cruiser known to the General Board during a hearing on regular cruiser design later in November 1930. "When it comes to the airplane I have always been very much in favor of using them to every extent that it [sic] could be used and advocate very much the experimental type of flying-on cruiser. If it works out all right it will revise warfare." Pratt clearly thought the idea had revolutionary potential and wanted to build as much tonnage of this type as the London Treaty would permit. Admiral Bristol suggested that the concept be tested by the War College in concert with other non-flying deck cruiser designs. The assignment of this responsibility naturally appealed to Pratt, a former War College president. Admiral Laning was also at the November cruiser hearing and reported to the General Board that the College had already attempted to game the flying deck concept, a reflection of his earlier coordination with Bristol and the Board. Laning reported that "The class unfortunately was not sufficiently trained to use that cruiser as I thought they would." Laning added that he hoped to see "one of those [flying deck cruisers] built, because it looks like it will have a good deal of value." Bristol closed the discussion by emphasizing that the Board looked forward to the design proposals of the bureaus and the results of the War College gaming.[62]

The General Board held three formal hearings on the proposed designs for the flying deck cruiser over the course of December 1930. During these hearings the flying deck concept met the cold, hard compromises involved in warship design. The Board's first order of business was to look at the design recommendations of the various bureaus. Admiral Bristol emphasized that the design would have to involve considerable compromises in order to achieve the characteristics of both a cruiser and an aircraft carrier: "This, as you all know, is a new idea in ships for the Navy and we have gone far enough in the Bureaus to realize that there are a great many features in this type of ship to be considered. . . . I would like to have each one of you as the hearing goes on keep in mind that the items enumerated are only a guide and I would appreciate the suggestion or remark of any one that would assist to develop this type of ship."

First to testify was Captain Van Keuren of BuC&R who presented the initial design for a cruiser with two triple-mount 6-inch gun turrets forward, one in the back, and a flight deck with island roughly amidships (see Table 2). The size of the ship was extraordinary, almost 650 feet in length and the design used up every last ton of the 10,000 allowed by the London Treaty. Van Keuren freely admitted that "In fact we may be a little squeezed on weights in order to get all that everyone wants and naturally it will be a compromise and something will have to suffer."[63]

A particularly innovative feature in the design of the ship was the correction to a drawback in the design of *Saratoga* and *Lexington*. These ships had been designed with their island, the above-deck tower and navigation bridge on their starboard sides. However, their flight decks had been aligned with the center-line of the ship, which was fine for the old *Langley* which did not have an island or above-deck gun turrets. However, on the two converted battle cruisers this was not done, in part to accommodate fire directors for the 8-inch guns. This caused a weight imbalance that was only corrected by leaving critical fuel storage empty on the starboard side to avoid a starboard list. Van Keuren's design adjusted for this by moving the flight deck off center to the port side of the vessel to counter-balance the island. This design innovation is still in the U.S. Navy aircraft carriers today.[64]

Van Keuren's design came under criticism from the BuAer representatives who then presented their counter-design (see Table 3). They wanted to remove the island completely because it interfered with flight operations. Aviators had little experience with an island at that point and so were still convinced that these structures degraded the flexibility of the ship to conduct flight operations. Most of their experience was with the *Langley*—a flush deck design from which they could take off from the bow or stern, a factor that meant the ship need not always seek to sail into the wind before launching its aircraft. Also, lack of an island allowed more planes to be stored on deck and thus increased the aircraft-carrying capacity of the vessel.[65]

The aviators were in turn criticized by the conservative Admiral Clark, head of OpNav's Fleet Training Division, and some of the BuOrd officers for neglecting the ship's cruiser characteristics. By eliminating the island the aviators would seriously degrade the fire control directors' lines of sight and fields of view for the 6- and 5-inch guns. The aviators, with the support of Admiral Pratt countered that technology might solve the fire control problem by integrating the directors into the design of the turrets. BuOrd's representatives seemed skeptical about this solution. This discussion highlighted Bristol's and Van Keuren's prediction that compromises would be necessary. However, no one seemed willing to compromise in his area of interest on that day.[66] These technical discussions led to a more basic issue, the operational role of the ship in the fleet. The aviators, principally through the testimony of Commander Turner and Lieutenant Commander Nicholson, proposed that the ship's main battery was really its aircraft. Turner also made the startling statement that he regarded the 5-inch anti-air (AA) guns as excessive, believing that heavy machine guns were more effective against dive-bombing

TABLE 2. Initial Design Proposal for Flying Deck Cruiser by the Bureau of Construction and Repair

Characteristic	Specification	Notes
Displacement (tons)	10,000 tons	
Length/beam/draft (feet)	630/65/20	
Deck Type/length*	Flush with island/ Length not specified but between 300-400 feet	BuC&R included option for eliminating the island for a "pure" flush deck configuration.
Maximum Speed (knots)	32.5 kts	
Aviation Facilities	–Open bay Hangar Deck –Flight deck One elevator forward One set of arresting gear	To reduce fire hazards Both flying on and off Aft of the elevator
Shaft Horse Power	80,000	
Radius of Action	10,000 nm at 15 kts	nm = nautical mile
Armament/Battery gun turrets	Primary—3 triple 6-inch to be portable Secondary—six 5-inch guns for anti-air/ .50 cal machine guns	Machine guns were assemblies.
Armor configuration."	"All or none does not refer to the "8-inch" protection around magazines. "6-inch" protection around machinery, turrets, and barbettes	"8-inch" protection thickness of the armor. It refers to protection against 8-inch shell fire at long ranges. 6-inch protection refers to protection against fire at intermediate to shorter ranges from other light cruisers.
Aircraft	18/34 (hangar vs. hangar and flight deck stowage)	Fighter (VF) and Scout (VO)/equal numbers of each.

*Note: BuC&R defined "flush" during this hearing as: "No fixed obstructions above the flight deck, no bridges, fire control arrangements and the stacks are of the hinged type [that fold out of the way]." BuC&R's flush flight deck proposal narrowed at its forward part to make room for anti-aircraft (AA) batteries.[67]

TABLE 3. Initial Design Proposal for Flying Deck Cruiser by the Bureau of Aeronautics[68]

Characteristic	Specification	Notes
Displacement (tons)*	10,000 tons	
Length/beam/draft (feet)*	630/65/20	
Deck Type/length	Flush with Navigation Bridge forward but level with flight deck/340 feet	BuAer eliminated the island and moved the after turret forward with the other two in their design.**
Maximum Speed (knots)	32.5 kts	
Aviation Facilities*	—Open bay Hangar Deck —Flight deck One elevator forward One set of arresting gear	To reduce fire hazards Both flying on and off Aft of the elevator
Shaft Horse Power*	80,000	
Radius of Action*	10,000 nm at 15 kts	nm = nautical mile
Armament/Battery*	Primary—3 triple 6-inch gun turrets Secondary—six 5-inch guns for anti-air/ .50 cal machine guns	 Machine guns were to be portable assemblies.
Armor*	"All or none configuration." "8-inch" protection around magazines. "6-inch" protection around machinery, turrets, and barbettes	"8-inch" protection does not refer to the thickness of the armor. It refers to protection against 8-inch shell fire at long ranges. 6-inch protection refers to protection against fire at intermediate to shorter ranges from other light cruisers.
Aircraft*	18/34	Fighter (VF)/Scout (VO)

*All these characteristics were roughly the same for the two designs.

**BuAer posited a second design that kept one turret forward and moved two aft to achieve the same air operations efficiencies inherent in a wider flight deck without an island.

aircraft. Five-inch guns would be ineffective, due to short reaction times, against level bombers. Eliminating the 5-inch guns would make additional weight available and greatly simplify both flight deck and fire control designs.

Captain Van Keuren, who had assisted Moffett with the initial design proposals for the flying deck cruiser at the London conference, addressed the issue of bomb damage. He had been one of the Navy's technical observers for all the bombing tests in the 1920s—from the German battleship *Ostfriesland* to the bombing tests on the U.S. battleships scrapped by the Washington Naval Treaty. Van Keuren emphasized that 500-pound bombs could, and probably would, cause considerable damage to a 10,000-ton treaty heavy cruiser. His arguments seemed to support the arguments of those who wanted to divest the ship of its 5-inch guns. Captain John Towers of BuAer suggested that if the flight deck was located forward with all the 6-inch guns aft, the ship could flee a superior enemy with its three rear-mounted turrets firing while the ship launched protective or attacking aircraft into the wind created by its 32-knot speed off the bow. If the enemy managed to close, it would face a double threat in rapid 6-inch gunfire and aerial bombardment, probably by dive-bombing.

The aviators were not united in support of Towers' testimony. Moffett remained very quiet, letting his talented subordinates—Towers, Nicholson, Turner, Kenneth Whiting, and Marc Mitscher—argue the merits of moving the flight deck forward or moving it more to the rear of the design. At one point Moffett did make a "general" proposal to include a folding smoke stack in order remove this obstruction during flight operations as well as to solve the problems that the hot gases and smoke caused during the recovery of aircraft. This innovative suggestion eventually became a feature of BuAer's design. Van Keuren had earlier dismissed this modification as weighing too much. When Admiral Moffett seemed to back down on the idea Admiral Bristol spoke up: "We want to develop the general idea." Throughout the hearing, Bristol was very involved—not typical for the Board's senior member—in soliciting and developing the myriad of ideas that were being brainstormed at this hearing—he wanted to ensure they were included in the transcripts so the Board could review them later if need be. By the end of the hearing it was clear that the Board needed another day of hearings since it had yet to hear the formal testimony of BuEng, BuOrd, and OpNav.

The first day's hearing had necessarily revolved around the carrier characteristics of the ship's design. On the second day the testimony was supposed to focus on those aspects of the ship that made it a cruiser. This was all-important because in order for the ship to have the veneer of legality under the treaty its utility and function as a cruiser must be manifest. Admiral Pratt was not just concerned about other countries lodging treaty complaints. Remember that Pratt's experience concerning what was achievable within the constraints of the treaties was considerable and also included congressional attitudes.[69]

The second day's hearing was exceptional because the Navy Sec. C. F. Adams personally attended, as did Assistant Secretary for Aeronautics Davis S. Ingalls, and the chiefs of the BuEng and BuOrd, Admirals Yarnell and Leahy, respectively. They wanted to ensure that their views were made known in the presence of both the CNO and secretary. They first looked at the design proposed by BuEng, which built on the conventional cruiser hearings of earlier that year (April and November 1930). BuEng had essentially fitted a smaller flying deck aft on the current design for a 10,000-ton heavy cruiser. It rapidly became apparent that this design was completely unsatisfactory as to its flying qualities: it was difficult to land on and could carry only 11 planes. Speaking on behalf of BuAer Captain Towers said "that was our idea for a pure 8-inch cruiser. We don't regard that as a flying deck cruiser." At this point Bristol steered the discussion away from the "pure" cruiser design back to an extended discussion of what the attendees foresaw as the primary operational mission of the new design. He allowed extensive discussion by the BuAer representatives, especially Commander Turner, on the types of missions the ship would perform as a cruiser. Clearly the theater of operation everyone had in mind for this ship was the Pacific. Turner argued that because of its air component, the flying deck design could outperform any other cruiser in convoy escort, raiding, and screening. He added that:

> As regards tactical uses: as a scout in the outer [screen] they should be very much better than any other type of cruiser. For inner screening against destroyers in subsidiary attacks they should be as good as any other 6-inch cruiser. And their work in an offensive screen to conceal the passage of a fleet should be important because they can keep fighting patrols in the air and stop enemy air attack or give warning of them. They should be very useful in support *of advance base operations* in obtaining preliminary information and making preliminary attacks. They can be used as *battle line carriers*. If the battle line carriers [at that time *Lexington* and *Saratoga*] have been destroyed one or more of these vessels can be diverted to that use. (emphasis mine)[70]

Notice that Turner emphasized the role the flying deck cruiser could play in assisting advanced base operations to support War Plan Orange and as a means to ameliorate the shortage of "battle line carriers."

All the other BuAer personnel concurred with Turner's comments. Commander Mitscher (of later World War II fame) added that there were "advantages of the smoke screen laying apparatus that can be attached to the planes in a ship of this nature for its own defense or for a possible attack against an enemy ship in screening that ship." Mitscher's comment brings to mind precisely these capabilities used by a combination of escort carriers and destroyers at the Battle of Samar off Leyte Gulf in 1944; actions that helped unnerve Admiral Kurita in command of a mighty force of battleships. After Turner and Mitscher finished, Admiral Pratt broke his silence: "I want to find out whether this ship is a cruiser." Pratt solicited remarks from BuOrd (Admiral Leahy) who posited that the 6-inch flying deck cruiser would have an "equal" chance of success against an 8-inch cruiser and a better than equal chance against a 6-inch gun ship one-on-one—startling testimony from the nominal head of "the gun club." Pratt immediately clarified his purpose for the Board: "I asked this particularly to draw out a point in case we should have any hearing in Congress about this not being a cruiser; that she could not put up a good gun fight. If you can show she is a cruiser regardless of her aviation facilities." Leahy backtracked a bit after Pratt's remarks, saying that in order to stand a chance in a head-to-head fight with another cruiser, the flying deck ship must have "equal fire control" facilities, which was not possible on a flush deck design.[71]

This give and take continued—Bristol taking particular care to find out if the aviators could accept an island structure (which would elevate the fire directors) with some loss to the flying characteristics. Bristol wanted to make sure that everyone's views were recorded in the transcripts. He was especially interested in getting the other admirals present—those not part of the General Board—to state their opinions as to whether the flying deck cruiser was "living up to the treaty." Leahy equivocated, saying that if it had a "full deck," which prevented elevated fire control, then it had "the characteristics of an airplane carrier." However, if the fire control was elevated he thought "one might very properly consider it a cruiser." Next up was Admiral Yarnell of BuEng: "There is no question whatever that if you build a flush deck ship with a lot of planes on board, whether she is an aircraft carrier or cruiser will be a debatable subject." However, Yarnell stated that if the fire control problem was solved the ship " . . . would be very valuable for all the purposes which a cruiser might be faced [sic]. . . . " Admiral Taylor from War Plans Division had a different take, saying that "Anyone who wants to stick to the letter of the [London] treaty with a deck for flying off and on, will say it is a violation of the Treaty."[72]

Perhaps the most outspoken critic was Admiral Clark, director of OpNav's Fleet Training Division: "Make the flying incidental and stick to a cruiser. I know the [naval] air force started from the first to make an aircraft carrier." His assessment was essentially correct, except he should have substituted "the Navy" for "air force." Admiral Clark was convinced that the aviators were simply trying to build more aircraft carriers using the cruiser tonnage. After the meeting he took the unusual step of writing a letter to the Board summarizing his views and had it inserted at the end of the hearing transcript. In particular he noted the irony between the aviators' position that level bombers could probably not be countered but that these ships should have a complement of fighters anyway. Clark was among those admirals in the Navy, not numerous, who were still skeptical about the efficacy of naval air power.[73]

The road ahead after the second day of hearings was clear. The Board needed more detailed designs to examine. BuC&R was tasked to draft several variations on the two design approaches. At one extreme would be a ship designed to have minimal cruiser characteristics and maximum aviation capability and at the other end the reverse. Designs were to include various compromise versions between the two extremes. Also, Admiral Pratt directed Admiral Laning of the Naval War College to conduct a combat comparison "between [an] 8" gun ship and 6" gun ship carrying [a] flying on deck." Pratt, intimate with the war gaming process from his days as president of the War College, provided detailed operational parameters for the tests: "The tests are to be made under all conditions of weather—rough sea and smooth sea; high visibility; low visibility; night and day. Also the relative values of each type will be tested and reported on in screen work—convoy work—commerce destruction, and relative vulnerability against destroyer attack at night."[74]

The final flying deck hearing of 1930 was held two days before Christmas. Most of the same personnel who had been at the previous hearing attended. The hearing's purpose was to provide firm design guidance to BuC&R in order to prepare a suitable blueprint for the vessel's construction. BuC&R brought in seven different designs for the Board to consider, designated with alpha-numerics (see Appendix 4). Of these designs the G and D-series (D, D1, and D2) elicited the most interest. The G-design was the most cruiser-like and the E and F designs were flush deck designs. The D-series were true compromise designs. They included an island, three forward 6-inch gun turrets, as well as an innovative design for a folding smoke stack on the starboard side. This last design feature resulted from a compromise. If the final design included an island, the operations of the deck could be compensated for by the temporary removal of the smoke stack, which also inhibited flying operations. Additionally, the removal of the smoke to the side decreased the turbulence caused by the hot gases and so improved landing performance by the pilots. The D-designs could carry the most aircraft, even with an island.[75]

By the end of the hearing it was clear that BuC&R would probably be directed to continue to refine the design for the ship based on one of the D-versions. Admirals Clark and Taylor remained skeptical of the concept. Taylor's last words on the topic were blunt, "I think you are designing a carrier disguised as a cruiser." Not surprisingly, Admiral Clark subscribed "to Admiral Taylor's remarks."[76] The holiday season did not prevent the Board from continuing to refine the design. On

December 26, the Board met with junior officer and enlisted personnel from OpNav and BuOrd about the ship's design and then forwarded a memorandum to the bureaus with the approved design guidance on December 30. The Chief of Naval Operations provided additional guidance in February after the General Board had held another design meeting in late January. The design template used was that of the D-series but had been renamed A2 (summarized in Table 4.)[77] Although there was significant skepticism about the new vessel, the General Board clearly pressed ahead with development despite the objections of some. It did this principally because of the active support and sponsorship by the chief of naval operations for the flying deck cruiser.

Meanwhile, Admiral Laning had set his War College students and faculty to work, designating the experimental ships as "Light Aircraft Carriers (CLVs)" for the gaming directed by the chief of naval operations. The design for the games used a series of engagements that compared two types of CLV with the performance of current conventional 8- and 6-inch cruiser designs—the CL-26 and CL-38 classes, respectively. Laning used two conceptual versions of the CLV. "CLV-6" had twenty-four aircraft and protection against 6-inch gunfire while "CLV-8" used the heavy cruiser design with twelve aircraft and 8-inch gunfire protection. Aircraft bombing effectiveness was based on bomb hit percentages from the fleet. These percentages were decreased by 30 percent to account for combat stress and AA self-defense. This correction was at odds with the War Plans Division, which had decremented bombing accuracy by as much as 50–75 percent in combat. The great difference can be accounted for due to BuOrd's skepticism of aviation bombing performance. The data included in the War Plans estimate had been added by Willis Lee, a former BuOrd staff officer.[78]

By early 1931, the initial results from the College's war games were available. The results were impressive. The CLV designs, especially the twenty-four-plane version with lighter armor, were decidedly superior to their counterparts in all mission areas: offensive screening, protective screening (guarding against "surprise attack"), defensive screening (against attacks by destroyers), long-range scouting, and convoy protection. The report added that "The CLV is a decided menace to any battle cruiser (or even battleship) that might be deployed in connection with enemy scouting forces." This was sure to get the attention of the more traditional officers in the Navy. The CLV concept had decided promise as a scouting force ship. This ship would be most important in early phases of an Orange War in locating the enemy and protecting the advance base operations of the fleet.[79]

One reason this capability was so urgent was due to War Plan Orange's requirement for long-range reconnaissance to locate ships of the Japanese Fleet. Orange requirements—reconnaissance ahead of the fleet by as much as 1,000 nm—could not yet be met by existing seaplane technology, although there was great hope under Moffett that the Navy's dirigibles might perform this function. Later the PBY-5 Catalina long-range patrol aircraft provided this capability—but in the early 1930s this technology had not yet emerged. It was clear that Laning and his cohorts at the Naval War College thought the innovative CLV might be the operational solution to this and a variety of other thorny problems.[80]

In June 1931, BuC&R's latest design was ready for reevaluation by the General Board. By this time, due to confusion with the light aircraft carrier designation as a "CVL," the flying deck cruiser had been redesignated by the CNO as a "CF." The refined design included one feature in particular that had received very little discussion during the hearings but was an outgrowth of them. During the December 1930 hearings, the Board had questioned the BuAer officers at length about launching and recovering aircraft on the shortened deck of some of the designs. The aviators had brought up the technique of taking off at an angle in order in order to avoid the island, or perhaps a forward superstructure, as well as to get a longer deck run. Evidently the BuC&R officers had paid close attention because they included a slightly angled deck in the final design. It was offset to the port (left) side of the ship in order to give the aviators more usable deck space for spotting (parking) and flight operations (see Figure 2 [from Zimm, 236]). When this innovation was combined with the folding smoke stack it achieved most of the characteristics the aviators had wanted from the original flush-deck designs.[81] The flying deck cruiser was the world's first angled-deck carrier design, an innovation that was a generation ahead of its time. Ironically, the British Admiralty would make the recommendations that caused the United States to convert the aircraft carrier *Antietam* into an angled-deck ship in 1952–1953. Angled decks made landing and taking off at the same time possible. They also vastly improved safety by allowing an aviator who missed the arresting wire to take off again. Had the U.S. Navy built this ship it would have probably discovered these advantages during flight operations and experimentation.[82]

In the meantime, the Board continued to press for allocation of the approved light cruiser dollars for application to the construction of the CF. In April they recommended that any funds that might be left over from the 1932 building program be used to lay down "one [cruiser] of ten thousand-tons with flight deck." During hearings on July 16 and 17, 1932 the General Board finalized its advice on this design for the secretary of the Navy. The Board was enthusiastic in its commitment to build at least one flying deck cruiser for prototype experimentation in the fleet—if not more—as soon as possible.[83] The secretary of the Navy and Chief of Naval Operations were in attendance for the first day of these hearings as was Admiral Rock, the head of BuC&R. Admiral Bristol presided over the hearing and opened the discussion by emphasizing that "The design of this ship introduces new and unconventional features not encountered in any design of a previous vessel. Although the ship has *excellent cruiser qualities* at the same time compromise has been made and *we go as far as possible toward improving the carrier factors.*" (emphasis mine) "Carrier factors" refers to the airplane carrying capacity and flight deck for this ship.[84]

Admiral Pratt suggested that the design be denuded of all its 5-inch guns in favor of machine guns for AA defense. He clearly agreed with Turner's assessments from the earlier hearing. This would solve some of its fire control, weight, and structural problems. The weight, especially, could be used to increase the armor protection, which would make the flying deck cruiser more "cruiser-like." However, there was some urgency in the discussion because of the opportunity that the Congress had offered in the budget to build this class of ship. Admiral Rock noted prior to the end of the hearing: "I would like to emphasize a little . . . about the speed of the building of this ship. Certainly we all remember how favorable all the members of the [Congressional Naval Affairs] Committee were for the flight-deck cruiser. It appeared to me that if we could get any ship through the Committee it would be a flight-deck cruiser and it may be that we might bring enough eloquence to bear to get the passage of the authorization and appropriation immediately and thereby gain several months. So we should really try to get these characteristics settled and get our design out so as to be entirely ready to place the contract without any delay. . . . It is such an important design."

Bristol promised BuC&R a decision by the Board "within forty-eight hours" and was as good as his word. The Board held another meeting, with reduced attendance, the next day and finalized the design recommendations for BuC&R.[85]

The urgency was not just due to the favorable congressional situation cited by Rock, but had to do with the treaty system as well. The Board was very concerned that the next conference on disarmament, scheduled for Geneva in 1932, might eliminate the flying deck cruiser as an allowed ship class, particularly if it had not already been built. There was also a draft proposal by the League of Nations to extend a building moratorium on all new naval construction until the completion of the Geneva Conference. There was the potential that the construction of the flying deck cruiser might be prohibited either specifically, or generally, due to an arms agreement in the near future. As it turned out the League of Nations adopted the resolution proposing a new construction moratorium that September. This had the effect of removing any possibility of beginning construction on the flying deck cruiser in 1932.[86]

The delay the League of Nations resolution imposed on the flying deck cruiser's construction was fatal. By 1932, the deepening depression had become the dominant factor in the approval of construction money for the flight deck cruiser. Simply put, how could the administration or Congress approve funds for this ship when other ships, already built, were being laid up and naval yards and stations

PROPOSED 10000 TON CRUISER WITH FLIGHT DECK

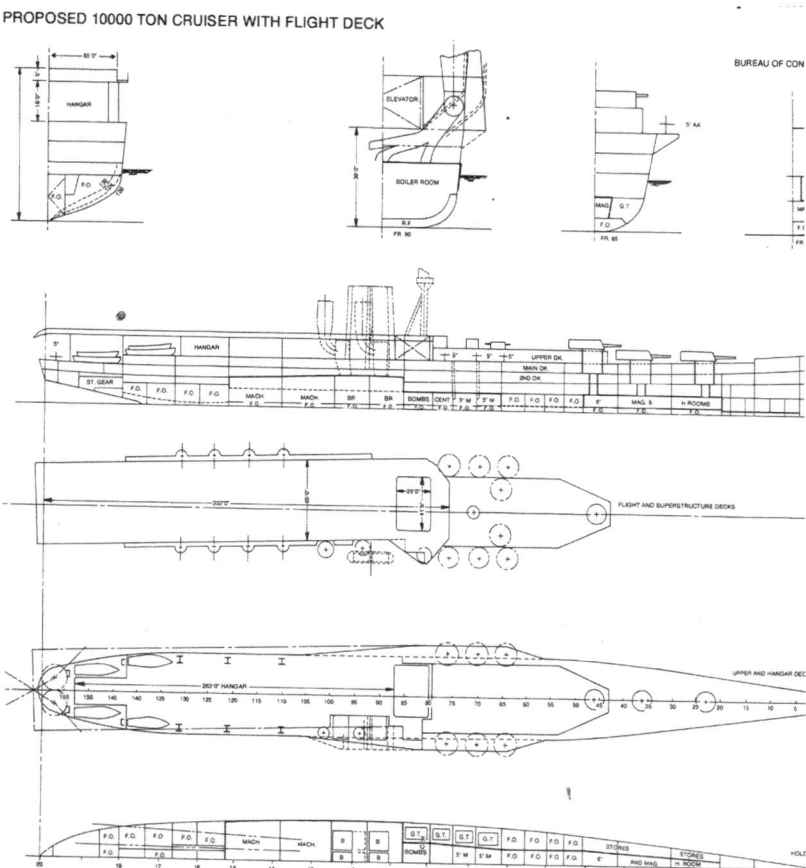

Length of Water Line 637' Beam on W.L. 625'
Displacement, Tons 11580 Draft 19.4'
Speed, Knots 32½ (Approx.)
Cruising Radius at 15 Knots, 10000 Miles.
S.H.P. at 32½ Knots, 80000
Type of Engines Turbines Mech Red. Gear
Number of Boiler Rooms Two, 3 Boilers each.
Main Battery, Guns 9-6" in Triple Turrets.
Main Battery, Torpedo Tubes None.
Torpedo Defense Battery 6-5" 25 Cal. A.A.
Aero Defense Guns 2—5" 25 Ca. A.A.
Main Side Belt Armor, Extreme Width 10'10"
Main Side Belt Armor, Depth Below W.L. 5' 0"

Main Side Belt Armor Thickness 110# over 30# S.T.S.
Barbettes, Thickness, Heavy Part 120#
Barbettes, Thickness, Light Part 120#
Turrets, Thickness: Port 200# Sides 70# Top 50# Rea
Conning Tower Proper, Thickness 60# Top 60#
Fire-Control Tower, Thickness ———
Conning Tower Tube, Thickness, Heavy Part ———
Conning Tower Tube, Thickness, Light Part ———
Uptake Protection Thickness ———
Protective Deck, Total Thickness over boilers & mach. 55#
Protective Deck, Total Thickness over mags, central, etc. 5!
Int. long. blks. mags, central, steering gear; 120# top, 50# b

PRELIMINARY DESIGN SECTION
Book No. 364   Date June 15, 1931 No. 011066

Figure 2. Flying Deck Cruiser Design Blueprint

TABLE 4. Characteristics for Flying Deck Cruiser by CNO and the General Board[87]

Characteristic	Specification	Notes
Displacement (tons)	10,000 tons	
Length (feet)	627	
Deck Type/length	Aft with island/350 feet	One elevator forward in center of flight deck
Maximum Speed (knots)	32.5 kts	
Aviation Facilities	—Open bay hangar deck	To reduce fire hazards
	—Flight deck	Both flying on and off
	One elevator forward	
	One set of arresting gear	Aft of the elevator
Shaft Horse Power	80,000	
Radius of Action	10,000 nm at 15 kts	nm = nautical mile
Armament/Battery	Primary—3 triple 6-inch gun turrets forward	
	Secondary—six 6 5-inch guns for anti-air/ .50 cal machine guns	Two 5-inch mounts aft. Machine guns were to be portable assemblies.
Armor	"6-inch" protection equal to that of a standard light cruiser	6-inch protection refers to protection against fire at intermediate to shorter ranges from other light cruisers.
Aircraft	24 VF and VO	Fighter (VF), Scout VO)

were being closed? Congressional hearings began for new construction requests in January 1932 for the Navy's plan to build two light cruisers, one of which was the CF. Admiral Pratt, savvy as ever to the prevailing winds of the budget environment, proposed that they at least build one vessel so it could be properly experimented with. Moffett, on the other hand, bluntly stated its real purpose—as a solution to the dearth of naval aviation in the fleet and as a platform that might well make all other cruisers obsolete: "By adding the flight deck cruiser to the fleet, carrier based aircraft will take their proper place with the main body (for the carrier is really and properly a capital ship) . . . at the same time aircraft will be provided for the all important duties connected with escort, patrol, blockade, scouting and raiding, for which the flying deck cruiser is so admirably suited."

Moffett proposed that all other naval appropriations give way in priority to the construction of the CF. The Committee gave the ship a favorable report. However, the Bill for its construction (only a single ship) was not acted on during that session of Congress. By the winter of 1932–1933 the economic situation had worsened and proposals were even being made to put the battleship fleet into lay-up in order to save money. In late 1932, the flying deck cruiser was removed from the construction budget for 1933 due to the governmental economies necessitated by the Depression.[88]

The flying deck cruiser died a slow death. However, the concept had enough merit to remain a General Board priority but the promising design of 1931 was radically altered. First, Admiral Clark had taken issue with the way in which aircraft bombing accuracy had been tested during the 1930–1931 Naval War College war games. A new round of games was conducted at the college in 1932 that used more pessimistic bomb accuracy decrements. Understandably, the CF's performance decreased in comparison to regular cruisers. However, another factor perhaps had more effect. From the beginning critics of the flying deck cruiser had labeled it a "hybrid." Retired Adm. Hilary Jones went so far as to call the ship "a hermaphrodite—neither a real cruiser nor a real airplane carrier. It has all the weaknesses of both and none of the efficient characteristics of either." This language had been written during the summer of 1933 for a Navy League speech when the issue of building the ship had again come up as the General Board drafted its annual construction request for 1934. The Navy League speech was forwarded to the Board by the Navy secretary during the Board's preparation of another recommendation on the topic. The Navy League had been a vocal critic of the Hoover administration's naval construction policies. The new administration, especially FDR, who was perhaps more sensitive to this constituency, wanted to ensure that the League's concerns were addressed. Jones, and active duty admirals like Schofield and the aviator admiral J. M Reeves, believed that "pure" aircraft carriers were really what the fleet needed.[89] With FDR's election, and the subsequent loosening of the purse strings for additional carrier construction, much of the original impetus and logic for the construction of the flying deck cruiser lost its persuasiveness. Also, the flying deck cruiser had lost some of its influential advocates. In 1933, Moffett died in the *Akron* crash, Pratt departed as CNO, Laning moved on from the Naval War College and Admiral Bristol had left the General Board in mid-1932.[90]

The new and "improved" data from recent War College games were forwarded to the Board and it was decided to hold new design hearings for the CF. The latest design reflected the bias of the games and the times. BuC&R now proposed a twelve-plane design which no longer had an angled flight deck. The flight deck was again amidships with the island in the middle.[91] This design did not fare well during the hearings of 1934 and Admiral Clark, again testifying before the Board, emphasized that since the original proposal the aircraft carrier outlook had improved substantially with the laying down of the *Yorktown*-class carriers. Perhaps what really killed the flying deck concept was the building of the treaty aircraft carriers that the Navy had wanted all along. Parsimony still played a role, however. When Admiral King, now chief of BuAer, continued to press for the CF's construction during the new hearings, he was forced to admit that it was a "20 million dollar experiment." King continued to support the concept, both as BuAer and later as chairman of the General Board and acceded to its cancellation only in 1939 when it was permanently tabled in favor of conventional cruiser construction. Only the War Plans Division—which had initially opposed the 1931 design—voted to retain it, probably because they saw its value as an asset that could attack Japanese invasion transports in defense of the Philippines. Ironically, some of these conventional light cruiser hulls were later converted into light carriers that looked a lot like the original flush deck designs for the flying deck cruiser.[92]

* * *

In summary, Navy leaders saw aviation as an operational means to apply sea power in the Pacific within the constraints of the fortification clause. The strategies of the various Orange Plans could not succeed without at-sea air bases. Tenders, carriers, and flying deck cruisers fulfilled an operational function in this regard. Accordingly, the Navy investigated two principal solutions in the early years after the signing of the first Treaty. The first solution was to rapidly build up to treaty limits in aircraft carriers, experimenting the entire time to determine which designs held the most promise. With only the *Langley,* a slow ship whose experimental value was limited, the General Board and BuAer further advocated building up the aviation component of the Asiatic Fleet as much as allowable under the Treaty, principally through the use of a class of ship not limited by the Treaty—aircraft tenders. Frustration in these two areas led to a third concept that promised a solution—the flying deck cruiser.

The flying deck cruiser was also born out of the strategic frustration due to the fortification clause. As such, the CF's history serves as a key example of the interactions between the treaty system and innovation in naval design and doctrine. The experience reflects the gradually changing institutional temper of the U.S. Navy toward naval aviation during the period. More and more aviation came to be seen as a part of the solution to operational and strategic problems resulting from the fortification clause of the Washington Naval Treaty. The CF's history also highlights the role of individuals. The names of some of the actors in this story—Turner, King, Kinkaid, Mitscher, Towers, and Leahy—all went on to operational and strategic commands in World War II. It was no accident that their attitude toward sea power had become one that was committed to "blue water" power projection in the absence of bases. They had spent the interwar period thinking long and hard about how to solve the immense difficulties of a war in the vast expanses of the Pacific in the absence of bases. True fleets are composed of men and ships. As such these agents of innovation went on to become the "brain" of the fleet that defeated the Japanese. The U.S. Navy was changing. It was more "blue-water" and "air-minded" than it probably would have been in the absence of the treaty system, especially in the absence of the fortification clause. The flying deck cruiser also highlights the organizational interactions and processes involved in the design and testing of new concepts in the interwar U.S. Navy. These interactions provided the institutional setting in which innovative ideas were proposed, tested, and—in the case of the flying deck cruiser—discarded. However, one cannot say that the flying deck cruiser did not get a fair hearing. In fact, the process that led to the 1931 design seems to have produced a very promising vessel, one that may have been more than simply an aircraft carrier in disguise. It had real potential and the Navy leaders of the period have received little credit for sensing that it did. That it was not built was due to a variety of factors, some of which were beyond the control of the Navy, such as the Great Depression and the naval construction moratorium of 1932. The flying deck cruiser was a result of the surprising commitment of most of the Navy's top leadership to a fleet that included a substantial and powerful naval air component to help make up for the treaties' constraints. When the Soviet navy of the Cold War era built its first "aircraft carriers"—the *Kiev* class—they looked remarkably similar to the final flying deck cruiser design of 1930. The Soviet navy's designation, in fact, for these ships was TAKR, which in Russian translates as "Large Aircraft Carrying Cruiser."

All three programs—carriers, airplane tenders, and the CF—ran into substantial difficulties as a result of the dynamics of the interaction of the Washington Naval Treaty and Republican budget policies. The Republican administrations and Congresses immediately following World War I had hoped to reap a "peace dividend" as a result of victory in Europe.[93] The system of naval limitations—and hopefully reduction—was meant to be one way of reaping additional dividends over time. In the case of naval limitation the ends were often confused with the means. Because of these dynamics the introduction of both carriers and tenders was greatly slowed and the flying deck cruiser never built.

Chapter 7

Strategic Innovation in the Interwar U.S. Navy

Floating Dry Docks

If we are making war on the Atlantic we have all our docking facilities available.... Whereas, in the Pacific we want to push across the ocean. We can seize plenty of places where we can imagine ourselves laying down a dock. Therefore, in any case the only place for floating dock[s] is in the Pacific because there we have to make war a long way from home and possibly without an ally.
  Adm. W. L. Rodgers, December 3, 1923 General Board hearing

The plan to create the most efficacious U.S. Fleet possible after the implementation of a naval limitation regime in Washington 1922 involved the intersection of technological advancements, new strategic and material conditions conferred by the treaty system, and novel approaches for operating the fleet at extreme distances without secure or available bases. In addition to the Navy's focus on warships, the plan for a treaty fleet included the construction of a mobile base for use upon the outbreak of war. The mobile base project (MBP) was another outgrowth of the fortification clause that limited the United States to one small naval facility in Manila Bay at Cavite. This naval yard had no dry docks or repair facilities. These facilities were essential if the United States deployed its main fleet for any length of time in the Far East. The facilities were barely adequate to handle the limited number of submarines, destroyers, a seaplane tender, and two or three cruisers of the small U.S. Asiatic Fleet. One of the major components of the MBP was the inclusion of a building program for large, mobile floating dry docks. This program, nested within the overall mobile base, can be characterized as an innovation at the strategic level of war.

This solution required a massive industrial effort. The MBP eventually developed into an extensive readiness plan for an entire series of advanced bases for the Pacific Fleet. These bases could be established where needed and then moved again as the advance proceeded. Initially, it was hoped, they would be constructed in noncontested areas—or areas of little naval threat—in the southern Philippines or the Japanese Mandate Islands (the Marshall and Caroline Archipelagos). Eventually the Navy leadership concluded, at the prodding of the Marine Corps and chief of naval operation's (OpNav) War Plans Division, that to establish these bases it would probably have to seize terrain from a hostile and determined enemy. In any case, the Navy realized it might not have the time in war to generate the industrial infrastructure to build deployable dry docks and so proposed that it build the infrastructure for its advanced bases ahead of time, especially the dry docks.[1] Above all, these dry docks must be deployable and therefore would be towed or driven under their own power to their future locations in the far reaches of the Pacific. These ideas became the mobile base project that was a secret appendix to War Plan Orange. The importance of the plan is emphasized by the fact that the docks were treated as a sort of "secret weapon." This concept was already in place, with a considerable logistics and construction schedule to support it, by 1923.[2]

It was not due to serendipity, last-minute planning, or accident that the U.S. approach to the Pacific War, which emerged in late 1941 and early 1942 was "logistics heavy." It could not be otherwise given the constraints of the fortification clause, the failure to build up Guam and Manila after the demise of the treaty system, and the interwar planning by the U.S. Navy. The massive infrastructure used to create a grid of mutually supporting Pacific bases in the South and Central Pacific had been planned in minute detail during the interwar period. Many of the contracts had already been written and some of the docks had already been built. Threat of war and the outbreak of war were the final ingredients that activated the readiness appendices of the war plans in earnest. These in turn helped bring the full weight of the United States' industrial might to bear.

No advanced bases existed beyond Hawaii prior to the Japanese attack in 1941. However, as we have seen, much of the planning and organization had already begun years before. The first of the great advance bases was established at Espiritu Santo in the northern New Hebrides Islands which was 500 nm southeast of Guadalcanal. Work was begun on this important base at nearly the same time that the campaign for the possession of an airfield on Guadalcanal took place in 1942. The first elements of the Espiritu Santo advanced base arrived in early July—barely a month before the First Marine Division landed at Tulagi and Guadalcanal. The number one priority was the construction of air fields, however, some PBY patrol seaplanes were able to operate almost immediately from an anchorage offshore. Forward deployed seaplane tenders like USS *Curtiss* serviced the patrol plane crews. By December, in no small measure due to this key logistics and air base, the Japanese were well on their way to defeat at Guadalcanal. They would never again have the operational initiative in the Pacific War. While the campaign was still under way, the Navy decided to make Espiritu Santo the first deployment site for the world's largest floating dry dock—Advanced Base Sectional Dock 1 (ABSD-1).[3]

The 1924 mobile base project served as a precursor to this gargantuan, war-winning logistical effort. The development of this plan also emphasized the evolutionary nature of innovation in the U.S. Navy during this period and its confluence with the treaty system. The need for seizing and establishing advanced bases had already been foreseen prior to the Washington Conference.[4] The Washington Naval Treaty, however, brought the issue of advanced base operations into the immediate foreground because of the fortification clause. A complete rewrite of the Orange Plan was required, including the addition of a brand new secret appendix dedicated to a mobile strategic base.

* * *

The mobile base project is among the little known stories of the interwar Navy. It was first reflected in the Navy's 1923 building policy. The highest priority naval construction program in the MBP consisted of massive, deployable, floating dry docks.[5] First a few words on dry docks. A land-based dry dock operates somewhat like the locks of a canal. A compartment is flooded and the level of water controlled to allow ships to enter and leave it. In the case of a dry dock, all of the water is pumped out so that essential maintenance work can be done on the hull, propulsion system and any other system (such as SONAR) located below the waterline. Floating dry docks have an additional requirement to give the dock enough buoyancy for itself to float as well as the ships being serviced. When naval officers refer to "self-docking" of these platforms it simply means that the dock can be made buoyant enough to access various portions of the hull without putting the dock itself into a larger dry dock. Just as aircraft carriers were the centerpiece for the operational solution afforded by naval aviation due to the lack of bases in the western Pacific, the floating dry dock was the centerpiece of the mobile basing solution.

Navy planners soon suspected that the fortification clause had compromised the original "thrusting" strategy of the Orange Plan. This strategy was contingent upon the expansion and fortification of naval facilities in both the Philippines and Guam.[6] With these naval facilities in place the U.S. Fleet could avoid a repetition of the fiasco that occurred during the Russo-Japanese War where the Russian Fleet had had limited logistics support in the theater of operations. The Japanese managed to bottle up one Russian squadron in its harbor at Port Arthur on the Manchurian Liaodong Peninsula. This squadron was eventually destroyed without its ever really getting out to sea again. A relieving Russian Fleet was dispatched to save Port Arthur but the vital fortress fell to the Japanese army before it arrived. This Russian Fleet was in turn destroyed as it attempted to reach Vladivostok at the key battle in the Tsushima Strait. A key element in the destruction of the Russian Fleet had been its deplorable material condition, in great part due to its lengthy cruise to the far eastern theater of operations. Marine life had fouled the bottoms of ships slowing them down and the boilers, through extended use, had taken a beating. The Japanese, on the other hand, had ready access to naval yards for repair and refit.[7]

The U.S. Navy had no intention of allowing a similar fate to befall them should war break out with Japan. The U.S. Fleet would "thrust" across the Pacific and base itself on Manila—where a smaller Asiatic Squadron (fleet) was already based. The similarity to the Russian situation at Port Arthur is remarkable.[8] Should the main battle occur en route, the U.S. Navy could fall back on its local support bases (such as Hawaii or even Guam) to lick its wounds and reconstitute.[9] Should the battle not occur, the U.S. Fleet would have at least secured its position in the theater and could seek out battle with its adversary after "trimming up" up its ships in the yards and docks of its western Pacific bases. Especially critical to the success of this plan was the role played by Guam, which it was assumed could be reached prior to the decisive clash with the Japanese Fleet. The final dash to Manila would occur from a fortified naval base at Guam. Applying the analogy of the Russian experience, the U.S. Navy would have two Port Arthurs instead of one. Additionally, as a third course of action, the planners added a plan to establish new bases in the southern Philippines against the possibility that Manila and the naval facilities at Cavite fell quickly to a Japanese invasion. This third option provided the basis of the more mature mobile base plan of 1924.[10]

With the loss of the possibility of fortifying Guam or improving the fortifications in the Philippines the entire Orange Plan had to be recast. It now became a plan with two courses of action. The first course of action acknowledged the probable loss of the Philippines and included extensive appendices for a "step-by-step" advance across the Pacific. The other course of action, which for a time held precedence, was a modified version of the thrusting plan that gave priority to the establishment of an interim base in the Marshall or Caroline Islands and a forward base in the southern Philippines, probably near or on the island of Mindanao (see Map 1). The closest the plan came to specifying an intermediate basing location was in a small archipelago southeast of Truk named the Mortlocks. Additionally, the Navy had to completely reorganize the fleet. This was reflected in Navy General Order 94 of late 1922, which reorganized the fleet into four component and mutually supporting fleets: the Battle Fleet, Scouting Fleet, Control Fleet, and Fleet Base Force.[11]

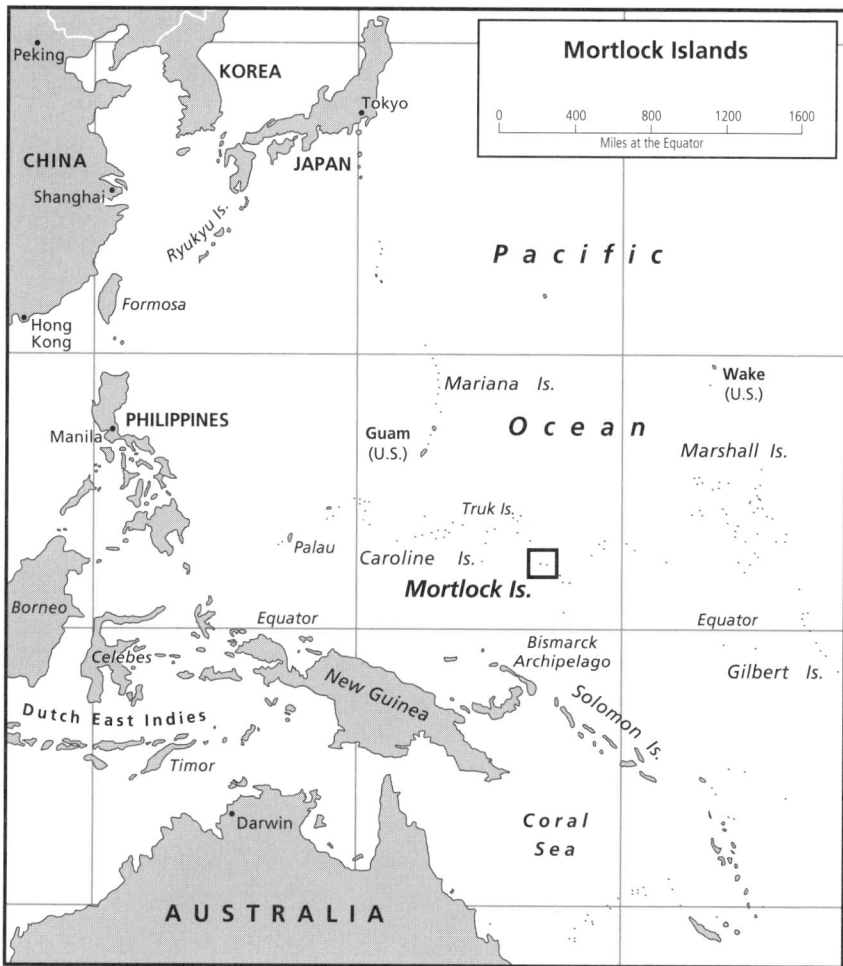

MAP 1. Map of the Pacific

The Fleet Base Force, which included the mobile base, played a major role in both courses of action. It could be used for either a deliberate advance, or audaciously pushed forward to establish an interim base in the Mortlock area, perhaps eventually even in the southern Philippines. Because of its utility in both courses of action, the MBP retained a very high priority for construction of its key component—large floating dry docks. This was reflected in the building serials of the General Board throughout the Treaty period. The status given to floating dry docks is reflected by their programmatic (although not fiscal) health as a building priority throughout the austere budget environment of the interwar period, especially that of the 1920s.[12]

Advanced Base Operations—The Requirement for a Mobile Base

A Navy civil engineer officer named A. C. Cunningham first broached the idea for a mobile base in 1904. The post-Spanish War expansion of the U.S. Navy under Theodore Roosevelt had brought the issue of naval logistics to the fore in the discussions and debates published by *The U.S. Naval Institute Proceedings*. Cunningham proposed a "movable base. The base was composed of sectionalized floating dry docks, colliers, ammunition, repair, supply and hospital ships [that] would move with or behind the fleet, and would offer all the essential service required of a base." However, the Navy did not pursue this idea at the time. It wanted to spend its limited dollars on ships and relied instead on existing continental bases for maintenance and support.[13]

At the time of Cunningham's proposal, the Navy already had an ambitious program under way to build two large floating dry docks—in fact, the first one was already finished. They were built in part due to the efforts of A. T. Mahan and Theodore Roosevelt, at first for use in the western hemisphere. The first one could accommodate a vessel of 18,000 tons displacement. It was already completed when Cunningham wrote his article and for almost 40 years was located in Algiers, Louisiana near the city of New Orleans. Interestingly, this dock was later designated YFD-2 and often was referred to as the "New Orleans dock" or sometimes just as "New Orleans." YFD-2 was towed out to Pearl Harbor in 1940 to increase that naval station's yard capabilities when the U.S. Pacific Fleet was moved to Hawaii on the eve of World War II. Her more illustrious sister was named *Dewey* and was a bit smaller, handling 16,000 tons of lift. Not long after her construction was complete, *Dewey* was towed across the Atlantic and through the Suez Canal in order to beef up the logistics and maintenance capability of the U.S. Navy's most important overseas base in Manila Bay. Both of these docks would see action in World War II; however, neither of them was designed for the type of mobile base operations envisaged by the mobile base project. Their moves to Pearl and Manila were one time only moves because of their poor seaworthiness and somewhat inflexible design. YFD-2, for example had had to be disassembled in order to get her through the Panama Canal in 1940.[14]

The idea for a mobile base resurfaced again in the newly established War Plans Division of OpNav shortly after World War I. The Treaty of Versailles had awarded Japan mandate over the former-German island groups in the Pacific. Of particular concern to the OpNav war planners were anchorages in the Marshall and Caroline groups that could provide havens for the operations of Japanese naval and air units against the flank of a U.S. Fleet advancing to Guam and then the Philippines. Recall that OpNav planners Lt. Cdr. Holloway H. Frost, Cdr. William S. Pye, and Capt. Harry E. Yarnell had addressed the need for advanced bases in an Office of Naval Intelligence (ONI) study of 1920, *The Conduct of an Oversea Naval Campaign*. This paper foresaw operations in Micronesia "to seize and occupy an advanced base." In the section of the study that addressed "the use of naval bases for the fleet" the authors brought up the issue of preparing a basing force in peacetime: "Nevertheless, in building up operating bases we should provide for some docking and repair facilities and plan for their rapid increase after war is declared." The authors gave specific numbers of bases desired for the campaign: two major bases for the "battle fleet," which implied locations at Guam and Manila, and at least seven "minor" bases for auxiliary forces in theater and along the lines of communication. Finally, the authors specified the key ingredient and listed the "chief" priority for the "supply and repair bases" as: "1. The towing of floating dry docks from the home bases to the repair base. It must be recognized that this is a tremendous undertaking. It will be impractical to build the docks after the beginning of the war, as too much time will be required. *The only means of meeting this problem seems to be to construct a number of floating docks* which will be used during peace at our home bases to supplement the fixed dry docks, and which are so constructed that they can be towed, *either intact or in sections* to the fleet bases." (emphasis mine) The study went on to specify the use of underwater ("marine") railways that could be used to repair smaller vessels such as destroyers and submarines, the requirements for ashore repair facilities, fuel and munitions storage, and material for traditional docking construction. The report also included the requirement for blueprints for "a large hospital."[15]

United States Marine Corps (USMC) Maj. Earl H. Ellis probably used this ONI study as a template for his "Advanced Base Force Operations in Micronesia" study of July 1921. The Marines, building on Ellis' work, tackled the problem of amphibious assault in the development of the "prescient" *Tentative Manual for Landing Operations*. This document provided the conceptual and doctrinal basis for the amphibious operations of the U.S. Marine Corps in World War II. According to historian Alan R. Millett, "The American forces that eventually defeated the Japanese were already well developed by the time the Japanese army began its China campaign in 1937. . . . " These initiatives were not a direct outgrowth of the fortification clause. Ellis' and ONI's studies on the requirement for amphibious forces and advanced bases had been completed prior to the Washington Treaty. However, the Marine Corps, OpNav, and the General Board were spurred down the path of advanced base, and ultimately mobile base, operations by the fortification clause.[16]

It was in the area of mobile base planning—the creation of deployable bases exclusive of the seizure of forward locations for the bases—where the fortification clause had its most direct impact. Historians have acknowledged the role, albeit briefly, of the fortification clause as the impetus for changes to the War Plan

Orange. Neglected, in most cases, was the original question: how does one fight in the Pacific without any permanent bases prior to the start of a war? For most of the interwar period this question—a direct result of the fortification clause—was germane to all of the Navy's logistic planning in support of the Orange Plans. Regardless of which Orange course of action was ascendant—thrusting or step-by-step—the essential logistics plan with regard to floating dry docks, transport (eventually to include amphibious ships), oilers, tenders, ammunition, and other types of auxiliary shipping remained remarkably stable. These were first reflected in the 1924 Orange document (WPL-9) and its secret mobile base appendix.[17]

The Navy was fortunate at the time of the Washington Conference to have as the head of its War Plans Division Adm. Clarence S. Williams, a General Board alumnus. Williams oversaw Orange planning from July 1921 to September 1922. He recognized the limitations of the thrusting course of action and had the good fortune to be in the right place at the right time when the Washington Treaty was implemented. It was Williams who instigated the rewriting of the Orange Plan, principally due to the impact of fortification clause on Pacific strategy. The clause was such a blow to the advocates of the "thrusting" advance that Williams was able to recast the second phase of the Orange Plan into the more cautionary step-by-step advance through Micronesia (although the thrusting course of action was still listed as an option). Williams' major changes were approved in late 1922 by Navy secretary Denby at which point a more detailed rewriting of the Orange Plan began.[18]

Williams' departure from the War Plans Division did not halt the momentum of the planning that resulted in the mature 1924 Orange Plan (WPL-9) with its mobile base appendix. Shortly after Williams' departure, the planners in OpNav formalized the idea of using a mobile base. In the words of Commander Coffey of the War Plans Division, "it became apparent that it was necessary to incorporate in [the war plans] what we have called a mobile base project to run along lines parallel to the shore establishment project." Coffey mildly rebuked the General Board during a December 1923 hearing on floating dry docks for waiting a year to act on "a written order" that directed them to determine the characteristics of the mobile base. Lacking guidance, the War Plans Division had "drawn up" its own "characteristics with regard to docks." Because this information was sensitive, it was not included in the transcript of the hearing, although Coffey identified "a definite number of docks and marine railways, two floating docks to handle the largest ships, two to handle 15,000-ton ships and twelve portable marine railways." Discussion during the hearing made it clear that the two large floating dry docks had already received endorsement for construction by the Board. This decision was also reflected in the Board's 1923 recommendation to begin construction on "one floating auto-mobile dry dock capable of docking . . . airplane carriers or any battleship" for fiscal 1926.[19]

Two major issues were discussed with respect to the floating dry dock program of the MBP. The first concerned using floating dry docks of a smaller type instead of twelve "portable" marine railways. The idea behind the marine railway was that ships could be positioned on underwater carriages and then moved on a rail line out of the water for work on the hull. Commander Coffey of the War Plans Division emphasized that this plan was a holdover from pre-Treaty war plans. The General Board was skeptical of the idea and in its questioning quickly determined that these "portable" railways were nothing of the sort. Once installed they were semipermanent and many members of the Board made it clear they preferred floating dry docks because they could be more rapidly redeployed (such as to a new island base) and afforded "a much greater choice of location...." Admiral Strauss, summarized the prevailing attitude: "Of course the marine railways are perfectly useless [in an Orange campaign] as you see now. The whole scheme is useless."[20]

Also discussed was the issue of the floating dry docks being self-propelled versus being simply towed to prospective anchorages. Here, too, the Board tended to prefer a solution that provided the maximum in strategic mobility and flexibility. Admiral Jones, in particular, thought the docks should be self-propelled, including the smaller ones. Coffey noted that the War Plans Division, if it got an auto-mobile dock, wanted "ten thousand miles radius" at twelve knots speed. The discussion also broached the idea of building the docks in sections in order to get them through the Panama Canal. The different sections could be built by a number of yards on both the Pacific and Atlantic coasts which would secure wider congressional support. Finally, the Board decided to let OpNav decide whether they thought self-propulsion was the preferred method of moving the docks.[21]

Coffey went back to the War Plans Division and incorporated these suggestions into the final version of the MBP. OpNav promulgated the plan as Appendix F to Plan Orange entitled "Basic Readiness Plan—Mobile Base Project" on December 20, 1923. A secret document, this plan had several other salient features. In its precept, the plan emphasized that: "It must be kept in mind that this 'Mobile Base Project' is a *readiness* plan prescribing the standards of readiness, as regards development of the project, to be *attained and maintained in peace*, and that it is not an *operating* plan. This Mobile Base will be incorporated in the U.S. Fleet Train (emphasis mine).[22]

The distinction between readiness and operating was important. By making this distinction the Navy hoped to justify peacetime spending for this capability. As events would show, the Navy would succeed in achieving its goal of building floating dry docks prior to the outbreak of war, although not on the scale outlined in the original MBP.

Secondly, Appendix F reflected the departure of Williams because it emphasized anew the use of the mobile base in the thrusting course of action. The purpose of the mobile base was to help establish the U.S. Fleet in "Manila Bay, to begin 14 days after Zero Day . . . not later than 60 days after Zero Day." This was an aggressive schedule indeed. The use of "temporary bases en route, for a period of one week" in the Marshall, Caroline, and other locations in the Philippine Islands was also specified as a design assumption for the project.[23]

The MBP now fully reflected the General Board's influence. The General Board recommended that the marine railways be eliminated in favor of a variety of floating dry docks which were now characterized as "mobile." The "mobile" characterization left the door open for either towed or self-propelled docks. The entirety of the docking facilities that were the centerpiece of the project was more ambitious than what was reflected in the original War Plans proposal. The following requirements were identified:

Docking Facilities
10. The assumptions in the preceding paragraph 7 (a) [numbers of ships and movements] will require the following docking facilities:

Class A   2 mobile floating drydocks to accommodate vessels above 20,000 tons displacement.
Class B   2 mobile floating drydocks to accommodate vessels from 12,000 to 20,000 tons displacement.
Class C   4 mobile floating drydocks to accommodate vessels from 3,000 to 12,000 tons displacement.
Class D   11 mobile floating drydocks to accommodate vessels below 3,000 tons displacement.

This section went on to give numbers of ships, expected periodicity for both emergency and routine docking, and anticipated sea space required for anchorage in square nautical miles. The total numbers of ships to be serviced was estimated (in the MBP) at 743 plus another 90 ships per month of naval transportation service (NTS, i.e., U.S. Army shipping). These were industrial-sized numbers, reflective of comprehensive planning, recent experience in World War I, and the impact of the fortification plan.[24]

The creation and inclusion of the mobile base appendix in the 1924 Navy Plan Orange can be characterized as evolutionary innovation at the strategic level of war. The ONI had commissioned studies by Frost et al. and Marine Major Ellis. These studies reflect that logistic planning was already underway prior to the Washington Conference. The war planning after the conference merely reflected the impact of the fortification clause in highlighting the importance of mobile, sea-based logistics to support a naval war in the Far East. The fortification clause did not cause OpNav's planners to write a mobile base readiness plan, but it did reinforce their inclination to give it priority in both the appendix and the key language of the parent Orange plans during the interwar period. By 1935 the plans for advanced and mobile base were very mature—only their projected operational employment and geo-location had changed. They had assumed even more importance because now they were to be used to create advanced bases to support a "step by step" advance through the Japanese mandates, a far more realistic course of action than had existed in previous years.

Floating Dry Docks—Changing Attitudes and Fiscal Realities

It is one thing to have a readiness plan that requires the construction of expensive floating infrastructure like mobile floating dry docks in peace, it is quite another to actually build that infrastructure. Once the General Board's membership had accepted the innovative floating dry dock program on its merits, they pressed ahead with innovative suggestions of their own. During the numerous hearings on the topic—there were ten hearings on the floating dry docks between 1924 and 1936—the Board's members not only advanced some interesting suggestions and recommendations for the design of the dry docks, but the hearings also reflect their changing attitudes about the relationship of sea power and mobile or advanced bases. A new understanding emerged which, in the words of Adm. Hilary Jones, became "axiomatic" and integral to the Navy's institutional conception of sea power.[25]

The Board's members now addressed the details of design as well as budget strategies. The hearings conducted by the Board early in 1924 were the occasion for a number of innovative suggestions by the senior member, Admiral Rodgers. It was Rodgers who recommended building the docks in uniform sections. He was proposing a form of modular construction that would ease the passage of the docks through the Panama Canal. Rodgers wanted to use these docks, designed for deployment in an Orange War ". . . as local substitutes for permanent docks in home waters. That is something to be considered. We might perhaps get the jump on war by just that much." Later studies for the time needed to build the largest class A and B docks indicated that even in wartime it might take as long as eighteen to twenty months to build and deliver these behemoths.[26] Such a delay weighed heavily on the minds of naval planners and the members of the General Board. These docks could be built in peacetime and then used as conventional docks until needed for deployment as components of mobile bases. Rodgers was strongly supported in this suggestion by Admiral L.E. Gregory of the BuY&D who added that ". . . if we wait until war is actually started by the time they [the floating dry docks] are ready for use I am afraid the war would be well under way." As head of BuY&D, Gregory, not the Bureau of Construction and Repair (BuC&R), would be responsible for maintenance and upkeep of these docks and was the design authority for this program. Gregory was also the Navy's liaison with the all-important Joint Munitions Board, which allocated material for the consumption of strategic resources used by both the services such as iron and steel.[27]

Rodgers and the other senior members of the Board—especially Admirals Jones and Strauss—were also clearly committed to a program that would build at least some of the more important docks, especially the larger ones, in peacetime. At one point during the hearing Admiral Jones emphasized the importance of strategic considerations in the design of the docks: "There may be some give and take somewhere but as this is a war measure, the strategy element is the one to consider." Jones was supported by a BuY&D engineer who emphasized the three critical design factors as "engineering, strategy, and cost."[28]

Rodgers' proposal had operational, budgetary, and political advantages in addition to its strategic merits. Rodgers' proposal for sectional docks, especially for the largest docks but not limited to them, had operational merit because it provided a level of robustness to the program for both movement and defense. First, sectional docks were a requirement for the largest docks in order to get them through the dimensionally constrained locks of the Panama Canal. Previously YFD-1, the *Dewey*, had been built in sections in order to get it through the Suez Canal and out to the Philippines to dock cruisers and the large auxiliaries (which potentially included the smaller aircraft carriers allowed for under the Washington Naval Treaty).[29] *Dewey* had not been designed in equal sections, but rather with a large middle section and smaller bow and aft sections. The design had been mostly driven by the need to tow it through the Panama Canal, however it was decided to tow her through the Suez Canal in order to avoid disassembly. When viewed in this light the floating dry docks for the mobile base plan were an evolutionary development of a preexisting design. Rodgers proposed sections that were all of the same size. Not only could they move through the Canal, but if one section was lost during towing or due to enemy action the rest could still be assembled and used for smaller vessels. Thus modular construction spread the risk since the sections would be towed individually to their new location for assemblage. As Rodgers put it, "the whole dock is not useless—you are better insured." Not only that but the dock could be reassembled to use as a number of smaller docks, thus allowing more dockage for smaller ships if the relatively fewer bigger ships were not in for a yard period.[30]

From a fiscal standpoint, the docks could be constructed one section at a time, which would spread the high cost of the largest dock (estimated at 6.5 million dollars) over a number of years. Alternatively, building the sections simultaneously at multiple sites offered the potential to build these docks more rapidly, getting a product that could be used immediately in the fleet once the first sections were completed. Also, by building a number of similar sections, construction could be spread out around the United States at shipyards on both the East and West Coasts, which would likely garner a broader base of political support. Admiral Jones, in particular, emphasized the need to be able to spread the construction contracts to include the East Coast of the United States. In this way, the Navy could obtain the maximum amount of political support for the project by distributing the contracts to a number of congressional districts. This provided another reason the docks must be able to transit the Panama Canal.[31]

During hearings on January 3, 1925, an exchange occurred between Admiral Jones and Captain Schofield that highlights changes in the Navy's institutional attitude toward sea power in response to the design of these floating docks for the treaty fleet:

> Admiral Jones: Isn't it almost axiomatic now that a mobile facility that will accomplish the same object is better than a fixed one?
>
> Captain Schofield: It is true; particularly now.
>
> Admiral Jones: Particularly anything from a military standpoint.

Jones and Schofield were referring to the situation occasioned by this hearing—the design of floating dry docks to account for the fortification clause. There was not a single dissent reflected in the transcript to their mutual agreement on this new "axiom."[32]

Rodgers' suggestion for modular construction was investigated by BuC&R's engineers. However, its 6.5 million dollar price tag for just the largest dock (Class A) caused the administration—and not a few Navy officers—to balk.[33] This amount was fully as much as the request for the first phase of the battleship modernization program and for only one unit versus the six battleships in that request. Nevertheless, the enthusiasm of the Board was so great that BuC&R went ahead with a detailed feasibility study. By the time BuC&R finished, estimates of the cost of the dock had increased to over 9 million dollars. The secretary of the Navy sent the Board and BuC&R back to the drawing board. It is a measure of the importance of this program that the General Board immediately reconvened in March of 1925 to consider a counterproposal by BuY&D for a new design for the floating dry dock program.[34]

BuY&D put forward a completely new design as an alternative to the expensive sectional floating dry dock. The first problem had been that when BuY&D looked at the sectional dock for the largest type, its cost estimates proved to be way off the mark, over 9 million dollars for just one Class A dock of fairly deep draft.[35] The cost was high due to the weight occasioned by the additional steel and design complexity needed to strengthen the joints between the sections. As an alternate course of action, the design engineer of BuC&R, J. Michaelson, proposed a completely new dock that came to be known as the "shipshape gate type" which was estimated to cost around 2.7 million dollars.[36] It incorporated a number of innovative design concepts that drastically lessened the cost and made the draft of the dock shallower. Shallow draft was an especially desirable characteristic given that the Class A docks were to be the central feature of the mobile/advanced bases being planned for in the Orange Plan. Another desirable feature of this design was that it could be used for the entire series of mobile base floating dry docks—Classes A through D. The smaller docks could dock in the bigger docks for maintenance. The outstanding problems with this design involved the fact that the largest docks (potentially to include the Class B used for cruisers and smaller carriers) could not "self dock"—that is they could not be ballasted so that their complete hull bottoms could be cleaned and maintained. This defect limited their useful life, especially if they were to be moved very often. Michaelson's new pointed-bow "gate" design for these docks incorporated a feature that later became a standard feature on U.S. Navy amphibious shipping—a floodable well-deck.[37]

During these hearings additional innovative features incorporated by Michaelson were highlighted. For example, the concerns over the risk now incurred by a unitary dock caused Michaelson to propose, and Admiral Gregory to endorse, an additional feature that would have allowed the largest dry dock to accommodate a wooden flight deck for protective air coverage during its transit. Additionally, the wooden deck could be disassembled upon arrival at the advanced base location and the deck lumber used to construct shore facilities. Admiral Jones closed the March 25, 1925 hearing joking that Admiral Moffett would use this feature of the dock to increase BuAer's requirement for more aircraft.[38]

The Washington Naval Treaty was specifically mentioned in the BuC&R's cover letter for the new dock design. BuY&D claimed that "Self-propulsion is not advocated as it would immediately place this dock in a class which would be considered a ship and therefore in contravention of the Treaty." Admiral Strauss of the Board immediately corrected this view: "It's the other way around. . . . I mean to say if it [the self-propelled floating dry dock] is a ship they can't count it as an installation. I take it we wouldn't have a right to send a 40,000-ton dock out to Manila and install it, but if you come steaming in with another ship to take a wounded ship out they couldn't object."[39] Obviously the "they" referred to were the Japanese. This exchange emphasizes the focus of the Board's members on ways to get around the fortification clause of the Washington Naval Treaty.

Another hearing was held in April 1925 after BuY&D had had time to consult with the design engineers of BuC&R. The Board made the recommendation to go with Michaelson's "gated" design. The Board also decided to reduce the size of the Class B dock to accommodate ships of 18,000 tons or less, which it thought would enable the smaller dock to transit the Panama Canal. The two bureaus also jointly raised the issue of building a smaller Class D dock first in order to experiment with the design in the fleet. The fundamental assumption that the largest docks needed to be built in peace was also included in the General Board's building policy recommendation in April 1925. The next year, the General Board specified in its January 1926 building recommendation that it preferred to see the Class A dock built first and this was eventually the type of dock included in the mature building plan for 1929. The Board was emphatic in recommending its construction arguing that it would be "useful in peace and indispensable immediately upon the outbreak of war."[40]

Ironically, the Navy did not build the largest dock first but instead built the experimental Class D dock. This outcome is consistent with the Navy's gradual approach, which was also a function of limited funds and stingy administrations during the treaty period. The first reason the larger dock was not built, despite its continued central role in the war plans, was the outstanding issue of "self-docking." The Chief of Naval Operations, Adm. C. F. Hughes, agonized over the Board's recommendation to go with the cheaper, riskier "gated" dock and finally decided in late 1927 to endorse the Board's recommendation that "Class 'A' and Class 'B' floating dry docks should be of the ship-shaped design. . . . "[41] The secretary of the Navy agreed but the construction of any floating dry docks was deferred due to the critical economic situation, the need for cruisers and carriers, and the Navy's decision after the London Conference to focus on only the most important warship construction programs to rectify the now fictional 5–3 ratio with Japan. It was not until 1931 that the Chief of Naval Operations, Admiral Pratt managed to get authorization for a smaller Class D type that could dock 1,500-ton destroyers and submarines. This dock was eventually designated ARD-1. This modest program was allocated funds and the Navy at last had its first mobile floating dry dock built in peacetime specifically for the mobile base. President Hoover may not have laid down any ships, but at least one floating dry dock was built during his parsimonious administration. Large floating dry docks, however, would have to wait for the election of FDR.[42]

By the time the mid-1930s rolled around, the war plans requirements for floating dry docks had gone through a number of changes. By 1927, in recognition of the difficulties associated with building any of these docks in peacetime, the War Plans Division had scaled back the mobile base project's nineteen docks to sixteen docks as follows:

One Class A dock (one), largest ship type—*Lexington*
  [a 36,000-ton aircraft carrier],
Two Class B docks (one), largest ship type—battleships,
Four Class C-1 docks (four), largest ship type—light cruisers,
Nine Class D docks (four), largest ship type—destroyers.

In the margins of the study these numbers had been reduced to ten (as reflected in the parenthesis above), this possibly reflects OpNav's budget strategy to build the first echelon of these docks. The C-1 designation reflects the elimination of a separate dock design for ships smaller than light cruisers to simplify the program and to allow the C-1 dock to undergo maintenance inside the larger Class A dock.[43]

The election of FDR and the subsequent availability of funds through the National Industrial Recovery Act (NIRA) revived the Navy's attempts to build the larger Class A and B floating dry docks. It appears the impetus within the Navy to again push ahead on the construction program came principally from three quarters: the General Board, the War Plans Division, and the fleet. Evidently the commander of the submarine force (Admiral Cole) forwarded a design (apparently unaware of the original design) for a large sectional floating dry dock to the then commander-in-chief of the U.S. Fleet Adm. Joseph M. Reeves in February 1935. Reeves then forwarded this proposal to secretary of the Navy. The General Board was consulted and in September 1935 instituted hearings to construct a modified Class A dock to be designated ARD-3. Similarly, in January of that year, the General Board had already held design meetings about building a second Class D dock of a bit larger design to be designated ARD-2. The momentum for building these docks in peace, for immediate use as part of the mobile base in war, was back on track.[44]

By 1937, only one small mobile floating dry dock, ARD-1, had been built and joined the fleet. The pressure to finish the construction on the others had increased significantly. The demise of the treaty system had not affected the Navy's plans for the mobile base. Admiral W. S. Pye, one of the original authors of *The Conduct of an Oversea Naval Campaign*, was now the head of the War Plans Division. He emphasized to the Chief of Naval Operations, Admiral Standley, that the floating dry docks would "perhaps be the deciding or least an important factor in obtaining a favorable decision in an Orange War." Remember that Pye had emphasized as a coauthor of the ONI study that "the only means of meeting this problem seems to be to construct a number of floating docks . . . " in order to project power across the Pacific. Standley evidently agreed with this logic and his interest in the topic spurred another study in which he had the War Plans Division compare the advantages of the sectional dock versus the ship-shape, gated dock as the final design for the largest docks. OpNav eventually recommended that the ship-shape gate type dock design for the Class A type be built—for all the original reasons cited by the General Board in 1925.[45] Earlier, the General Board had re-examined the issue of self-propulsion for the larger dock. These docks were now regarded almost on the level of a secret weapon. The Board's final hearings in 1935 on the topic of self-propulsion closed with a recommendation to include the shaft and mountings for self-propulsion but if discovered, put out the cover story that this was needed simply "to get the dock to Pearl Harbor" and avoid "spreading it out in the technical papers of the world that this is a mobile dock. . . . "[46]

The Chief of Naval Operations, now Adm. William Leahy, submitted the Navy's final opinion on the matter of the dock in a February 1937 statement for the Navy secretary. He emphasized the extremely high priority the Navy gave this program, pointing out that in 1935 the Class A and D docks (ARD-3 and ARD-2) were "items 4 and 5 in a list of six hundred and ninety projects which constituted the development program approved in 1935." In justifying this expensive construction he identified that this course of action was a result of "The Washington Treaty [which] has prevented the development of adequate fortified bases with repair and docking facilities in Guam or the Philippines." Immediately following this justification, Leahy reflected the new attitude that the U.S. Navy had developed: "In this matter of mobility we have developed *in a way which is our own.* We have more than any other nation developed our 'Mobile Base.' We have repair ships, tenders, store ships, hospital ships, refrigerator ships, etc., etc., which are a part of the fleet, known as the Fleet Base Force. . . . *Therefore, we ask you to let us go ahead with our ideas of making the U.S. Fleet always more mobile*; even if our floating dry docks are not agreeable to those who think only in terms of fixed defenses . . . in view of the considerations advanced above, we believe the additional increase in cost per dock is money well spent." (emphasis mine)

Leahy's remarks capture the change in the institutional conception of sea power that had occurred during the life of the floating dry dock program. It was a gradual, but powerful, transformation. It came from one of the architects of victory in World War II—Admiral Leahy—during the period of the treaty fleet and the fortification clause.[47]

Despite the rhetoric of the new Chief of Naval Operations, the construction of docks lagged behind requirements. By May 1940, BuY&D could only list three floating dry docks in the Pacific. These were the very old *Dewey* in the Philippines, ARD-1 in San Diego, and the deployment of *Dewey's* sister-dock (YFD-2) from New Orleans. YFD-2 had been sent to beef up the docking facilities of Pearl Harbor in October 1941 as a stop gap until ARD-3 could be completed.[48]

Ironically, elements from Leahy's appeal and Rodgers' initial proposal for a large sectional dock were adopted. Leahy's request for the smaller ARD-2 class docks of a unitary ship-shape design resulted in the most ambitious floating dry dock construction program of all time. The construction programs were well under way before World War II. For example, ARD-4 was laid down prior to Pearl Harbor in September 1941. With war, the Pacific Fleet would see these docks deployed to Dutch Harbor in the Aleutians (ARD-6), Midway (ARD-8), Fremantle, Australia (ARD-10), and Milne Bay in New Guinea (ARD-9).[49] As for the ARD-3 program, it was cancelled in favor of Rodgers' original 1925 proposal, now supported by Admiral Reeves and others. This dry dock was named in accordance with the original concepts broached by the General Board and OpNav—the Advanced Base Sectional Dock or ABSD-1. This dock was constructed almost exactly per Admirals Jones' and Rodgers' suggestions in sections at multiple shipyards. ABSD-1 (later renamed *Artisan* AFDB-1) was built in ten sections at four different shipyards on the East and West Coasts. Her East Coast sections were towed through the Panama Canal and the dock and its crew were commissioned formally in May 1943. Originally designed in ten sections, this massive dock was rated at being able to hold a 90,000-ton load (today's nuclear powered aircraft carriers are 95,000 tons). As Rodgers suggested, the dock could still be used if one section was lost. This actually occurred when one of the sections was lost in Assisi Bay near the New Hebrides Port of Espiritu Santo. Nevertheless, the nine remaining sections were joined and the dock was immediately put into service as the centerpiece of the advanced base off of Espiritu Santo. As foreseen, the dock did not remain fixed but was instead disassembled and towed again to the Philippines just as the war was ending. ABSD-1 was not the only such behemoth but rather the lead unit of an entire class of ABSDs (numbers two through six) to see service in the Pacific during World War II.[50]

Prior to the start of the Second World War the Navy had three floating dry docks of almost 42,000 tons capacity (YFD-1, YFD-2, and ARD-1) in service.[51] Despite the loss of *Dewey* (in the Philippines) and the damage to YFD-2 on December 7, 1941 by the Japanese surprise attack, by the end of 1941, the Navy's floating dry dock capacity had more than doubled to over 108,000 tons. This figure is evidence of a very robust construction and reflects a program that had been conceptualized during the interwar period and whose construction began before the war began. By the end of 1943, Navy floating dry docks in all theaters had increased their capacity to over 720,000 tons. By the end of the war in the Pacific, the Navy had over 400 advance bases established throughout the Pacific serviced by 152 floating dry docks of all sizes—from the lowly ARD-1 to the mighty ABSD-1.[52] A truly remarkable transformation had taken place. However, it was not due to war alone that the Navy made this shift. The Navy was intellectually and programmatically prepared for the scope of the challenge of war in vast reaches of the Pacific, a challenge it had conceptualized in great measure due to the constraints of the fortification clause of the Washington Naval Treaty.

* * *

Despite's Leahy's and the General Board's attempts to keep the "mobile base" a secret, the Japanese were well-informed. Japan had recognized the threat posed by this strategic initiative early on in its naval general staff plans to counter an American advance to relieve the Philippines. The base force is accurately shown in Japanese documents from the 1920s steering a direct course for the great anchorage at Truk in the Carolines.[53] Japanese tacticians ignored this threat; for them the mobile base did not represent a problem. For the Japanese the defeat of the main body of the U.S. Fleet—the battleship line—would obviate anything that the base force might do.

The Navy did not make a radical change to its paradigm of sea power during the interwar period. Instead it tried to adjust its solutions and designs to fit an existing paradigm or conception of sea power that was rooted in the teachings of A. T. Mahan. In doing this, the Navy changed the boundaries, but not the essentials, of the paradigm. Fleets, maritime commerce, and basing were all still regarded as essential to the sea power equation. However, the definition of just what "basing" really meant had been expanded to include mobile bases. Both fixed and mobile bases were needed in the strategic environment of the Pacific. Although the paradigm did not change, the fleet did. This change was accelerated by the anomaly caused by the fortification clause to the Navy's traditional conception of sea power. It became axiomatic that in wartime mobility was essential, not just of tactical and operational units but for strategic logistical capabilities. This change was vital. Commitment to the mobility of strategic assets remains a core competency of the U.S. Navy today.[54]

Chapter 8

Perspectives from Great Britain, Japan, and Germany

> ... *An absolute unmixed benefit to mankind, which carried no seeds of future misfortune.*
>
> Arthur Balfour on the Washington Treaties

> *As far as I am concerned, war with America starts now. We'll get our revenge, by God.*
>
> Adm. Kato Kanji at the Washington Conference

The treaty navies of Great Britain and Japan went in different directions than the U.S. Navy, which was consumed by the problem of the fortification clause of the Washington Naval Treaty. Japan and Great Britain were also limited by the constraints of the fortification clause, but for a variety of reasons this article did not have the major impact on strategy and design for them that it did in the United States. For Japan, the overwhelming factor that affected innovation was the inferior position the naval 5–5–3 ratio conferred upon it vis-à-vis the other two major naval powers—first at Washington and later at London in 1930. In Great Britain, the treaty system's influence on innovation was reflected in the relationship between naval arms limitation and a succession of government policies that linked security to reduced naval expenditures. In addition, the German navy (*Kriegsmarine*) will be examined given that it, more than any other post–World War I Navy, was limited by treaties as to the fleet it was allowed to retain and build. Comparing the experience of these other three navies with that of the U.S. Navy illuminates further the latter's experience with innovation. It also serves to highlight further the important role that naval arms limitation had—both positive and negative—on innovation in general during the interwar period. Comparison brings the experience of the U.S. Navy—its successes and shortcomings—into clearer focus.

* * *

The historical starting point for any narrative involving the factors behind the naval treaty system of the interwar period is the Treaty of Versailles, the so-called "hard peace" signed after World War I but not ratified by the U.S. Senate. The "war to end all wars" resulted in a treaty whose purpose was not to so much to end war but rather to ensure the security of its victors. Keeping this in mind, there are two principal areas of subsequent naval developments that Versailles impacted. First, Versailles must be included in any naval discussion because of its profound impact on innovation in both Germany and Great Britain. Second, Versailles left in place an unstable security structure in the Pacific that led directly to the Washington Treaties.

The Treaty of Versailles was imposed on a defeated and war-weary Germany. The severe constraints this treaty placed upon the *Kriegsmarine* were a direct outgrowth of the fearsome toll taken by German submarines in the last two years of World War I as well as the serious challenge posed by Germany's High Seas (surface) Fleet. The Imperial German Navy (*Reichsmarine*) sank more ships in World War I than during World War II, over a much shorter period, and with fewer submarines.[1] Also, the existence of a credible German dreadnought battleship "fleet in being" imposed a terrific strain on Great Britain's maintenance of maritime supremacy in the North Sea. Versailles' principal naval features addressed these points: submarines were prohibited for Germany to build or possess and the once mighty High Seas Battle Fleet was to be reduced to the status of a coastal defense navy. Additionally, the new German navy, as with the entire German military, was forbidden to develop aviation because of the new technology's fearsome potential as an offensive weapon.

Versailles drastically limited Germany's surface navy. The Germans' fleet of modern capital ships had been the source of a pre-war naval arms race and a scare at Jutland during the war. According to the terms of Versailles, Germany was allowed no more than six battleships of 10,000 tons (the tonnage limit for other nations' heavy cruisers later at Washington) with a maximum caliber of 11-inch guns for the main battery. They were also limited to six light cruisers of no more than 6,000 tons with 6-inch guns. Finally, Germany's navy was rounded out by twelve small torpedo boats. Further limitations on personnel were meant to keep Germany from effectively manning even these limited numbers of ships.[2] The

Kriegsmarine was not allowed to build any submarines, aircraft, or aircraft carriers. The German High Seas Fleet, which was sequestered by the Allies at Scapa Flow, had already disappeared—scuttled by its sailors before the British could take ownership of it. By 1919 German sea power was no more.

The prostrate Germans were allowed a naval force barely adequate for what British naval theorist Julian Corbett characterized as a "fleet in being"[3] and this fleet had yet to be built. Obviously such a humbling of the mighty led to a collective sigh of relief by the British political establishment. Historian John Terraine stated the matter best: "The nub of the matter was 'the total destruction of German seapower,' without that central fact firmly in mind it is not possible to make sense of much of what followed during the next twenty years." The Treaty of Versailles set the stage for a fiscal environment in Great Britain that favored economy for the Royal Navy (RN) over realistic assessments of what threats might develop in the future. This attitude existed for practically the entire interwar period.[4]

The second area of impact of Versailles was its influence on the balance of power between the United States, Great Britain, and Japan in the Far East. Prior to World War I, the pillar of Pacific security was the Anglo-Japanese naval alliance. This relationship, and Japan's hunger for Germany's Chinese and Pacific possessions, brought Japan into the war on the side of the Triple Entente.[5] Ironically, Japan's participation in World War I destabilized the Pacific due to Versailles' formal recognition of Japan's mandate under the League of Nations over the former German Island chains: the Marianas (except for Guam), the Carolines, and the Marshalls. Because of this imbalance Colonel House, President Wilson's representative at Versailles, ". . . . fought what was then referred to as the 'Naval Battle of Paris.'"[6] The United States wanted an end to the Anglo-Japanese pact while the British wanted to retain global naval supremacy. In the end, Britain agreed to support the League of Nations if the United States settled for less than naval parity—effectively a renunciation of the "navy second to none" policy of 1916. When the United States Senate refused to ratify Versailles, the naval deal between the two countries fell through. Great Britain's alliance with Japan remained an unsettled issue. Rumors of Britain's intent to renew the alliance with Japan in part prompted the Harding administration to initiate the invitation to the major powers to attend a conference in Washington to discuss disarmament and other security issues in 1921 (see Chapter 3). Versailles, therefore, must be the point of departure in any discussion of the treaty system involving Great Britain, Germany, and Japan. It contributed to the chain of events leading to Washington, effectively eliminated Germany as a naval power, and tacitly recognized the preeminence of Japan, the United States, and Great Britain as the first-rank naval powers of that day.[7]

Great Britain

English military historian Correlli Barnett calls the Washington Treaties "one of the major catastrophes of English history." In addition to criticizing Great Britain's penny-pinching governments, historians also accuse the British of the same sin as their German, U.S., and Japanese counterparts—a supposedly misguided Mahanian dedication to the capital ship and decisive fleet action. This anti-treaty argument identifies the Washington Naval Treaty and the Four Power Treaty (which nullified the Anglo-Japanese naval alliance) as the culprits for the "catastrophe." These treaties gave Great Britain a false sense of security that the "Jutlands and Trafalgars" of the future would be fought somewhere east of Singapore between the American and Japanese Fleets and therefore not involve too great an investment of British naval shipping.[8] This argument is also used to explain the minimal impact of innovations such as ASDIC (sonar) during the interwar period and the convoy tactics developed during World War I. That these innovations did not hold primacy within British doctrine was in part due to the altered strategic context imparted by the treaty system and its impact on the British psyche.

British attitudes toward the efficacy of air power and their classic strategy of using continental allies to fight continental opponents were also factors affecting interwar naval innovation. The French army emerged in 1919 with an enhanced reputation, especially given the disarmament of Germany. France could deal with any security problems on the continent of Europe. Additionally, the inflated claims of the air power enthusiasts led to a belief by the British political establishment that the Royal Air Force was at least an equal of the Royal Navy in defending or deterring threats against the British Isles.[9] Although not strictly treaty-related factors, both air power attitudes and strategy contributed to the strategic sense of security, first established after Versailles, which undermined British preparation for war from 1919 to 1939.

British attitudes toward the Americans and the Japanese also contributed to this false sense of security. These attitudes were also influenced by the treaty system. Singapore was not covered by the fortification clause and the Washington Treaty established U.S. preimminence in capital ships over the Japanese. The British profoundly underestimated the capabilities of the Imperial Japanese navy due to racially derived misconceptions. They also overestimated their own position in the Far East, especially the presumed invulnerability of their naval base at Singapore. Also, the British tended, given their attitude toward the Japanese Fleet, to overestimate the power of the U.S. Fleet to constrain the Japanese in the event of a crisis in the Pacific.[10]

British naval innovation is examined in two primary areas: submarine warfare and naval aviation. In both areas the British have received substantial criticism.[11] However, when one examines the criticism, the treaty system is only one of many, and often not the primary, factors that impeded naval innovation. The British thought the threat of the submarine had been abolished at Versailles. The subsequent London Treaties only further reinforced British governments' tendency to adopt false economies at the expense of naval development, tactics, and modernization to face the submarine threat. After the 1935 London Treaty removed the fundamental basis for these policies the British remained wedded to the idea that ASDIC (sonar) would be their "silver bullet" to defeat any potential future submarine threat.[12] The low esteem in which British officers held antisubmarine warfare (ASW) also ensured that neither the best nor the brightest in their officer corps were developing the necessary tactics to counter the threat. Nor were the required escorts being programmed for construction to complement ASDIC's promise as an ASW tool. War in 1939 found the British woefully unprepared for the undersea threat, even for the very small number of operational submarines that the Nazis managed to deploy.[13]

In the area of ASW, the impact of the treaty system was perhaps most visible. The treaty system, especially if we include Versailles within its parameters, was thought to have eliminated the threat of unrestricted submarine warfare against commerce. The course of the Treaty agreements reflected the continued anxiety of British civilians about the submarine and proposals were offered at every major conference to outlaw the submarine. The failure of these proposals highlights the disinterest of the Royal Navy in the submarine threat. Ironically, the Royal Navy continued to build submarines of its own during the interwar period. With the Anglo-German Naval Treaty of 1935, an appeasement-inspired agreement, and the subsequent breakdown of relations between Britain and Germany, the submarine loomed again as a threat. However, the evidence suggests that the Royal Navy's attitude had become hardened during the treaty period into one of neglect as to the efficacy of their ASW doctrine. Instead of focusing on a renewed submarine threat, and preparing accordingly, the British relied on ASDIC to counter any problems with submarines.

With the submarine menace supposedly eliminated convoy doctrine was relegated to the past. If the submarine did emerge again as a threat it could be hunted with ASDIC-equipped ships and killed—a return to the anticommerce raiding doctrine of the past. Convoys might not be needed again. However, this view was maintained within the overall context of post-Versailles agreements (including the Washington Treaties) that had legislated unrestricted submarine warfare out of existence. Submarines might threaten, but mostly warships and troopships, not neutral merchant shipping as they had during World War I.

In the case of the development of British naval aviation, the treaty system played a secondary role when compared to doctrinal and organizational factors. Air power theory and theorists, and their perverse influence on the lines of responsibility for naval aviation in the British military, not the treaty system per se, delayed and defused British development of naval aviation doctrine and equipment. The creation of the Royal Air Force resulted from the unification of air power under a central organization. Air power advocates such as Hugh Trenchard and Billy Mitchell argued that air power was best employed in mass. These arguments extended to organizational structures and resulted in "dual control" over the naval aviators and programs of the Royal Navy by the new Air Ministry. Historians have argued that this led to a neglect of the world's foremost naval aviation force, that belonging to the Royal Navy at the end of World War I, during the fiscally constrained interwar period. Especially damaging in the Royal Navy, as compared to the situation in United States and Japan, was the lack of flag officer patronage or the development of a flag officer path for naval aviators. This led to a neglect of naval aviation personnel and greatly diminished the esteem with which naval aviation was held within the naval service. The U.S. Navy's creation of a Bureau of Aeronautics headed by Admiral Moffett under the Navy department, and the organizational support of naval aviation by the General Board and OpNav, created a much firmer foundation for the development of naval aviation as compared to the Royal Navy.[14]

Another organizational factor that was perhaps in play has to do with the organization of the executive level leadership of the Royal Navy. Like the United States, the Royal Navy was under civilian control, in this case the Admiralty which was led by the First Lord of the Admiralty, a political appointee. Underneath the First Lord was the service chief, the First Sea Lord, who was somewhat analogous to the U.S. chief of naval operations, but far more powerful. There was no real counterpart in the British system to the General Board. Strategic policy within the Royal Navy was not subject to the sort of collegial give and take that we see occurring at the General Board hearings and in the dynamics between the bureaus, the Naval War College, and OpNav. In short, the executive level organizational leadership in the Royal Navy was far more centralized than in the U.S. Navy.

The role that the treaty system played in influencing innovation in the area of British naval aviation was important, but not all-encompassing. The treaty system, especially the dissolution of the *Reichsmarine*, simply provided a background of security that tended to undermine the arguments of naval officers that their naval aviation branch was inadequate to meet operational and strategic requirements. Also, Britain's lead in existing carrier tonnage at the time of the original Washington Naval Treaty ironically prevented the design and construction of new carriers and even aircraft in the fiscally constrained environment of the interwar period. It was difficult to get the civilians to buy "replacement" vessels and aircraft with a large operational aircraft carrier inventory already on hand.[15]

The British also developed a mobile base plan. This, however, was not due to the treaty system, but due to the strategic situation in the Mediterranean. British mobile base planning prepared them mentally, if not materially, for World War II. However, it had nothing to do with the treaty system and everything to do with the challenge posed by Italian naval and air, and later combined Axis, forces in this theater critical to British interests. The British were aware that it was essential they be able to generate sea power via forward basing in an austere theater. Historian Stephen Roskill identifies the planning and doctrine that were written for the Mobile Naval Base Defense Organization (MNDBO) as providing the basis for the design of equipment later developed for the Normandy landings in 1944.[16]

The treaty system kept Royal Navy officers focused on traditional platforms. The battleship remained the measure of naval power and the cruisers (including battle cruisers) remained the essential tool for control of Great Britain's extensive imperial sea lines of communication. The treaties did nothing to change that focus within the Royal Navy and instead, as the Washington and London conferences show, tended to emphasize rather than deemphasize this traditional focus. Political realities that limited military spending influenced the course naval innovation in a war-weary, and later economically depressed, Great Britain during 1922–1938. The treaty system provided a sense of strategic security that was used to realize a "peace dividend" through minimal naval expenditures. Also, British leaders felt that most threats could be deterred by the existing Royal Navy in concert with the Royal Air Force, the United States Navy, and the Army of the French Republic.

Japan

How the Japanese perceived the treaties and how they felt affected by them were conditioned by their views about sea power. Japanese sea power doctrine was overwhelmingly a product of their close study, and indeed "reverence," for the teachings of A. T. Mahan. Their focus, however, centered—as it did to a lesser degree in the other major naval powers of the period—on the single decisive battle. All of the Imperial Japanese Navy's (IJN) energies, and much of their innovation, can be explained by this doctrinal focus. However, their dedication to the doctrine of decisive battle, when combined with the advent of the treaty system, ended up exercising a malignant influence on innovation. Indeed, recent historians argue that the IJN's "Faith in the decisive battle doctrine became dogma in the Japanese navy."[17]

The Japanese experience during the interwar period was in some ways similar to that of the British. Civilian governments in Japan favored naval reductions and economies for the same reasons as their counterparts in Great Britain and the United States. Japanese governments became particularly stingy after the great 1923 Tokyo earthquake (which destroyed several warships). They were already predisposed to reduce naval budgets before that in the face of an ongoing naval arms race with the United States, which most of them realized they could not win. However, the legacy of the treaties for Japan was a serious fracturing of civil-military and military-military relations. In the case of the navy this resulted in the marginalization of the traditionally moderate officers—the so-called "treaty faction"—who were Kato Tomosaburo's successors at the naval ministry.[18] The militant "fleet faction," led by the firebrand Adm. Kato Kanji of the Imperial naval general staff (a separate organization from the naval ministry), soon gained ascendancy within the IJN after the elder Kato's death in 1923. Kato Kanji was driven by the "unbearable humiliation" he and others felt at having the 5–3 ratio imposed upon the IJN and favored negotiations that recognized Japan's right to naval parity with the United States.[19]

Japanese naval doctrine developers during the period made no significant operational departures from the "decisive battle" concept that was the fundamental basis for everything else.[20] The Japanese innovated, but for the wrong type of war. Instead of building the sort of navy one might expect for an extended naval war of attrition, the kind the United States might fight, the Japanese developed a navy for a short, decisive, but limited war.[21] The "get even" attitude of the Kato Kanji clique contributed substantially to a focus on avoiding, at all costs, an inferior position with regard to naval auxiliaries once the supposed damage had been done at Washington in 1922. This attitude bedeviled the remainder of the conferences dedicated to limiting naval armaments, especially the London Conference of 1930. The desire of the militants to build to treaty limits, as well as unconstrained production of naval auxiliaries, conflicted with their civilian counterparts' diplomatic efforts to increase stability while reducing naval expenditures, particularly after the beginning of the worldwide depression in 1929. After the ratification of the London agreements in 1930, open conflict between the militarists and the civilian and naval moderates erupted in a series of assassinations that took the lives of two Japanese prime ministers.[22]

The inferior position perceived by the militant fleet faction led the Japanese to develop tactics to maximize the performance of their cruisers and destroyers which were unlimited by tonnage ratios until the 1930 compromise at London. The Japanese also took advantage of the serendipitous fallout of the original Washington Naval Treaty with respect to their aircraft carriers. Some have claimed that the Washington Treaty forced the Japanese and Americans to develop carriers as the centerpiece of their navies, but this was not quite the case. The unintended consequences were subtler. The conversion language of the Treaty had allowed Japan and the United States to each convert two almost-finished capital ships into aircraft carriers.[23] These big carriers—*Kaga* and *Akagi* for Japan and *Lexington* and *Saratoga* for the United States—allowed both countries to critically examine them against the performance of the smaller carriers built within the tonnage limits of Washington.[24] Both countries eventually came to the conclusion that the larger carriers were more flexible and the Japanese reflected this lesson in their design for the large fleet carriers *Zuikaku* and *Shokaku*. Also, the propulsion plants for these converted ships, which were designed to move with or ahead of the heavily armored battlewagons, gave them excellent speed that was a great asset in creating their own "wind" for the expeditious launching of aircraft.[25] In Japan, as in the United States, naval aviation was promoted by the treaty system and came to be seen as a means to ameliorate what Japan saw as its unjustly inferior position. However, naval aviation was not seen as replacing the classic battle line in fighting the decisive fleet action.

Like Great Britain, Japan ignored the proven dangerous potential of the submarine. The rush to forget the submarine threat was a direct result of Versailles and subsequent agreements that outlawed the unrestricted submarine warfare practiced by Germany in World War I.[26] Japan had provided ASW escorts for World War I convoys in the Mediterranean, so the arguments that attribute their failure to a lack of experience do not stand up to close scrutiny. Instead, Japanese naval views were similar to those of the British, who saw ASW as an inferior form of defensive naval warfare, merely an adjunct defensive necessity to protect the battle fleet. Japanese naval officers preferred instead the offensive roles of the more prestigious capital ships and carriers.

Japanese officers believed that defending the battle fleet was more warrior-like than defending the lowly merchant fleet, despite the merchant fleet's strategic value. As a result, ASW was given short shrift as a critical warfare area in the IJN. Little intellectual attention and few resources were committed to the defense of merchant ships against the submarine. The island empire that Japan eventually conquered at the outset of the Pacific War would actually have more need of a well-protected merchant marine than a battle line. The battle line might be the fist but the merchant marine was the circulatory system. Japan neglected the ways and means by which they would protect their excellent fleet of cargo ships and tankers. This neglect, too, can be laid more at the door of a commitment to decisive battle that was reinforced, rather than restrained, by Japan's response to the treaty system during the interwar period.[27] As for submarine offensive operations, Japan consistently opposed banning submarines but more from the standpoint for their use as fleet assets. Submarines would weaken the approaching U.S. Fleet as well as serve in the reconnaissance role. Their role as commerce raiders was barely considered during the interwar period, despite the extensive lines of communication, the U.S. Navy would have to protect in its own advance across the Pacific.[28]

It was in the area of training smaller surface combatants—cruisers and destroyers—and night tactics that the Japanese intended to make up for their perceived loss of ground in 1922. Commentators identify relentless Japanese tactical training as evidence of their intention for "using a few to conquer many."[29] Night attacks, sophisticated destroyer squadron group torpedo and gun tactics, development of long-range, reliable torpedoes (sometimes called the "long lance"), and longer-range and heavier gun armaments on their ships were all means to this end.[30] Adm. Kato Kanji, now commanding the Combined Fleet in the late 1920s, drove his subordinates relentlessly in night training exercises that practiced these tactics, risking collision and damage. These exercises were meant to validate ship designs, doctrine, and weapons in an effort to offset quantity with quality. They were so rigorous and realistic that on one occasion, in 1927, a collision of several ships (cruisers and destroyers) resulted in over 120 casualties. Admiral Kato simply declared that this was the cost to be expected if they were to defeat the U.S. Navy. To their regret, the Allies learned in the opening year of the Pacific War about these tactics in the Java Sea (February 1942) and during multiple engagements around Guadalcanal. Savo Island and Tassafaronga (two of the more spectacular tactical naval defeats in U.S. Navy history—excepting, of course, Pearl Harbor) were a direct outgrowth of Japanese naval tactical, technical, and doctrinal innovation inspired by the hated "10:10:6" ratio.[31]

The fortification clause of the Washington Treaty led the Japanese to undervalue certain tactics and operational approaches while overvaluing others in the development of their Navy. The result was that they built a navy that was unable to properly address the challenge posed by the logistics-heavy and methodical amphibious operations of the U.S. Navy. The IJN, as mentioned, was designed for a short war, not a long duration war protecting a far-flung maritime empire. The Japanese case serves to highlight further the different approaches taken by the U.S. Navy in response to the treaties, especially the fortification clause. The Japanese were less dependent on logistics than the Americans and developed more along tactical lines. The United States, on the other hand was more dependent on logistics to begin with and more so once the treaty regime was in place. The Japanese organizational climate within their navy also serves to highlight some positive attributes of the U.S. system. In particular, the competition between the Imperial naval general staff and the naval ministry both staffed and led by active-duty naval officers, were the antithesis of the collegial and collaborative relationships between their American counterparts. Instead these organizations became the organizational breeding ground for two opposed navy "factions," with the "fleet faction" occupying the General Staff and the "treaty faction" ensconced in the Navy Ministry. Worse yet, it became more and more difficult during the interwar period for the Japanese Navy Minister to impose discipline on his nominal subordinates in both the fleet and the General Staff. This environment starkly contrasts to that within the U.S. Navy during the same timeframe.

Germany

In the German case, the influence of the Versailles and Second London treaties was profound. Undeterred by Versailles, and more likely spurred by it, Germany developed the most sophisticated submarine tactical and operational doctrine during the interwar period. Additionally, Grand Admiral Raeder's so-called Z-Plan for war with Great Britain was an innovative extrapolation on the commerce *guerre de course* the Germans knew they would have to wage against a superior British Fleet. Nevertheless, the 1935 Anglo-German Naval Treaty did limit the number of operational submarines should war break out early with Great Britain. Accordingly, it may have delayed certain technological innovations in that warfare area—especially since Hitler did not count Britain as Germany's enemy after that agreement and told his naval leaders as much.

Grand Admiral Erich Raeder was the head of the *Kriegsmarine* for most of the interwar period. He initially built the *Kriegsmarine* with France, Russia, and Poland in mind as the most likely naval adversaries. However, the evidence suggests that the treaty system only increased the Germans' desire for prohibited weapons and technology. In limiting German naval construction the treaties— especially Versailles—channeled Raeder, Karl Doenitz, and the German naval staff into conceptual approaches that resulted in more realistic and effective tactical and operational doctrine. In the case of the Z-Plan, the Germans achieved something rarer, a workable naval strategy to employ against Great Britain.[32]

The case of the German navy during the interwar period mirrors somewhat that of the German army. Versailles denied the German navy the ability to construct new weapons such as aircraft, submarines, and aircraft carriers as well as denying them the right to build dreadnought battleships. The German naval leadership was forced to think conceptually about a naval war despite this lack of resources. When political and strategic conditions finally allowed her to begin construction of a lesslimited navy, the result was not so much preparation for another Jutland as it was an effort toward a coherent naval strategy aimed at Britain's vital sea lines of communication. The case in Germany was very much one where drastic limitation combined with thoughtful consideration of post-war lessons resulted in a new menu of operational and strategic approaches—in short, innovation.[33]

Of primary importance, as in the case of Great Britain, was the Treaty of Versailles. By denying the Germans their most successful naval weapon of World War I, the Allies ensured a clandestine program of U-boat development—especially in design, tactics, command, and control. All the accounts rightly identify Karl Doenitz as the chief architect of innovation in this regard.[34] The key Allied response that checkmated the German campaign in World War I was the convoy system. Doenitz recognized this and spent a considerable level of effort developing group (*Rudel*) tactics to counter the convoy on the principle of meeting "concentration with concentration." In 1927, Doenitz was promoted to command a half-flotilla of torpedo boats that he essentially treated as a training opportunity to develop U-boat tactics. He especially practiced group night attacks with the goal of destroying a convoy. During the 1929 naval maneuvers his half-flotilla located and destroyed an "enemy" convoy in a night attack. Like the Japanese, Doenitz was relying on superior performance at night to outweigh the material and technological advances of his adversaries.[35]

However, until 1937 Doenitz also supported the naval expansion construction that included battleships. Doenitz claimed that he foresaw the war-winning potential of the submarine early on, but this claim is not supported by his actions. Doenitz's estrangement from Raeder's balanced-fleet approach in the Z-Plan occurred only after Britain and the Royal Navy had been identified as the most capable adversary in any future naval war. Recent evidence indicates that Doenitz, even at the war's end, was more supportive of Raeder's program than his memoirs attest.[36]

Doenitz had prepared intellectually and conceptually during the interwar period, despite a lack of actual submarines, for another submarine war. His four pillars for U-boat operations were: concentration against concentration (the future wolfpacks), night attacks, centralized command and control by Commander-in-Chief U-Boats (Doenitz ashore in Germany), and decentralized execution by the submarines upon contact with the convoy.[37] The imperative for reliable command and control for such a dispersed and difficult-to-reach force posed a significant problem, which Doenitz intended to solve with encrypted radio. The result was one of the World War II's most famous machines, the Enigma encryption device—a masterpiece of cryptologic innovation. However, Doenitz was at odds with the naval staff in this matter. The staff regarded this sort of command and control as a liability, believing that breaking radio silence would compromise U-Boat locations and survivability. Doenitz placed all his faith in Enigma and hindsight tells us that the naval staff was correct.[38]

The Anglo-German London Naval Conference was extremely important with regard to the German naval innovation. Doenitz' ideas were bankrolled by the Anglo-German Treaty's explicit approval for a German submarine construction program. The principal impetus for this conference was Hitler's renunciation of the terms of the Treaty of Versailles on March 16, 1935.[39] Great Britain regarded the new Hitler regime differently than the Germany they defeated in 1918. Germany was discounted as a credible naval threat. The imperatives of fiscal economy also resulted in Great Britain allowing Germany a considerable amount of latitude in rebuilding their navy should they choose to do so. Germany was allowed to build up to 35 percent of British surface tonnage and 45 percent in submarines. What is more amazing, and a key to how benignly British governments viewed Germany during this period, was the stipulation that Germany could build up to 100 percent of Britain's tonnage for submarines with special notification and at the expense of other classes of ships' tonnages. Thus Germany's submarine construction was sanctioned officially by treaty with Great Britain.[40] In sum, the naval treaties led to a focused program of submarine innovation by Germany. In denying the Germans the actual ability to build submarines the limitations instead focused the Germans on intellectual efforts that led to perhaps more profound innovation. Second, the London Treaty of 1935 enabled the Germans to build the submarines they had thought so long and hard about.

By 1938, it was apparent that Germany was indeed a potential threat to the peace of Europe and the security of Great Britain. Before long, Germany had abandoned all pretense of adhering to treaty limits and began construction of a balanced fleet (the Z-Plan). This plan's purpose was clearly to challenge the Royal Navy's ability to protect the British Isles' sea lines of communication. Some historians have argued that the construction of the German Fleet after the repudiation of Versailles suffered from an overly Mahanian focus on capital ship construction for decisive battle at the expense of the U-boat Fleet. They often cite Doenitz in making this judgment.[41] This view mirrors Doenitz' post-war criticisms, which lament the slow pace of submarine production prior to 1939. Historian John Terraine calls Grand Admiral Raeder's Z-Plan (the post–1937 naval expansion plan) "rubbish." Terraine goes on to question Raeder's intelligence in ignoring the lessons of World War I in initiating the building of fast battleships (prior to the 1935 London Treaty and the Z-Plan).[42]

Adm. Friedrich Ruge, who served on Rommel's staff during the war and later at the higher levels of command, offers an alternative explanation that seems more reliable. Ruge argues that Germany's fleet of surface ships was part of an overall plan for a balanced fleet to raid British commerce. It was not intended for the single decisive Mahanian sea battle, but rather this fleet's purpose was to more efficaciously attack Britain's vulnerable sea lines of communication (SLOCS) via combined arms action by surface, subsurface, and air forces. In early twenty-first-century terminology, the Z-Plan was an asymmetric approach to a naval war with Britain. Also, the Z-Plan was predicated on not fighting Great Britain until 1948 (at the earliest) and so was subverted not long after its implementation by Hitler's declaration of war in 1939.[43]

The Z-Plan was not a design for a Mahanian Fleet of battleships that would go head-to-head with the British battle line. Raeder envisioned a number of operational elements, all working together, that would slowly strangle British commerce while at the same time disputing the Royal Navy's command of the sea. The *Kriegsmarine* would consist of a home fleet, composed of older battleships and escorts that would tie down a commensurate or greater number of British warships close to home. Additionally, task groups built around fast battleships would raid from bases in the North Sea (and eventually Norway) to try and break up British convoys. Raeder anticipated that these groups would slowly whittle British capital strength down as submarines and aircraft picked off dispersed British merchant shipping. Finally, there would be another echelon of extremely long-range submarine and surface raiders that would tie down additional British naval assets in distant theaters such as the Mediterranean, Indian Ocean, and South Atlantic.[44] Raeder should rightfully be considered a key innovator for overseeing the design of this plan. The treaty system that provided the strategic context led to Raeder's strategic innovation as reflected by the Z-Plan.

Subsequent historical evidence supports the argument that Raeder never intended for his navy to fight one major sea battle but many, with the principal target being escorted convoys and choking Britain's maritime commerce. It was clear to Raeder and others that whatever they did, the *Kriegsmarine* would never outnumber the Royal Navy at sea. Raeder knew he must keep his fleet literally, as well as figuratively, "in being" for a war against British commerce and communication. The ships and submarines the Germans had upon the outbreak of war in 1939, despite their limited numbers, allowed the *Kriegsmarine* to sally forth far more than the High Seas Fleet of World War I. Surface, air, and submarine raiders would complement each other in attacking the vulnerable British sea lines of communication. To this end, Raeder designed his surface and subsurface raiders for speed, endurance, and survivability. The British were horrified when they realized how survivable the German commerce-raiding battleships were. Both *Graf Spee* and *Bismarck*, for example, were never sunk but scuttled to avoid capture despite considerable battle damage.

The Z-Plan included the construction of a number of aircraft carriers to be used against convoys and their covering forces. However, Germany's fledgling naval aviation was crippled from the start by Hermann Goering's appropriation of all aircraft design and operational control to himself and the Luftwaffe. Cooperation between the Navy and the Luftwaffe remained contentious throughout the war. Ruge offers the opinion that had Germany been allowed to build an aircraft carrier during the interwar period (instead of attempting to build one on the eve of war), they would have seen this weapon's value and produced more of them during the conflict. In the case of naval aviation, Versailles may have in fact suppressed innovation in an area of key importance and promise.[45]

Raeder's Z-Plan envisioned a fleet design that addressed the strategic problem directly, much in the way the *U.S. 1922 Naval Policy* addressed the fortification clause directly. However, the Z-Plan was never intended to be executed in 1939 and as the war wore on the resources Germany had to devote to its naval offensive against Britain's commerce, Germany became more and more reliant, as in World War I, on the single technology of the U-boat. Ultimately, the effect of the treaty system on Germany was a mixed bag. The treaties had forced the German naval leadership into a complete reappraisal of their view of strategy based on World War I. The shock of losing their navy may have had a positive impact on the ability of the German admirals to more realistically prepare for naval war with Great Britain. Had Hitler not initiated his war ahead of the Z-Plan's schedule, the innovative developments in the interwar German navy might have produced some spectacular, if tragic, results.

Comparing the U.S. and German navies of the period illuminates the role of constraint in influencing innovation. Both navies concentrated on long cruising radius, superior damage control capabilities which led to greater survivability, and attempted to employ the most sophisticated propulsion systems. The *Kriegsmarine* was the only other navy of the interwar period to approach (or exceed) the U.S. Navy in these categories. One explanation might be that the Germans also had a cultural affinity for technological and engineering solutions. However, the factor of treaty-related constraints established the basic strategic context that drove both nations to emphasize these design characteristics.

At the strategic level, both navies seemed to have a better view of the type of war in which they would probably be engaged in the future—an attritional war with significant material and logistical challenges. The Germans saw that their best naval strategy against Great Britain would be a war against Britain's commerce, a lesson they had not forgotten but learned well from their World War I experience. On the other hand, the Americans eventually came to the realization that a naval war in the Pacific with Japan would also be a lengthy affair that might involve years of methodical, and logistics-heavy, naval operations—a "hard slog" across the Pacific.

* * *

On the balance, the overall effect of the Treaties was not nearly as detrimental to naval readiness as some have claimed.[46] The German case is perhaps more pronounced in that both Raeder and Doenitz, to a great degree, innovated in the absence of a naval force structure denied them by Versailles and under considerable constraints even after the 1935 London Treaty (vying for resources with the other services). In any event, the Germans innovated where Britain was most vulnerable. Who is to say what might have been possible had Hitler waited until later, when the balanced fleet of the Z-Plan would have been more numerous in both surface ships and U-Boats and would have included a few aircraft carriers? Also, Hitler's invasion of the Soviet Union in 1941 guaranteed that the German naval offensive would remain under-resourced and a secondary priority for the rest of the war.[47]

In the British case, the Treaties lulled them into a false sense of security. Versailles reaffirmed Great Britain as the preeminent naval power but scattered the seeds for the severe challenges she faced at sea throughout World War II. A mere three years after Versailles the Royal Navy had been matched by Japan in the Far East and the United States had nominal parity. The United States would have had actual naval superiority had they finished building the modern ships whose keels had already been laid. After the Washington Conference, Britain's pillar of security for the Pacific, their alliance with Japan, was but a memory. The passing of the Anglo-Japanese naval partnership increased Japan's isolation vis-à-vis the West. Also, the worldwide peace movement, exemplified by the Kellogg-Briand Pact, when combined with "naval holidays" and their extensions, created a false ambience of comparative safety on the seas—particularly for Great Britain. S. E. Morison caustically referred to initiatives such as the Kellogg-Briand pact outlawing war as an "incantation." Unfortunately, most of Britain's interwar governments relied heavily on these incantations. Finally, this false sense of security combined with niggardly fiscal policies produced false economies in the Royal Navy which seriously jeopardized Britain's security when war broke out with Germany.

The Japanese experience was a mixture of both the good and the bad. The Japanese were perhaps unmatched by any other navy at the tactical level by the end of the interwar period. The Japanese also had the most veteran and experienced aircraft carrier force in the world. However, Japanese innovations resulting in a tactically elegant navy were negated by a flawed strategy—a direct result of an inferiority complex engendered by the process begun in Washington in 1922. Japanese adherence to a misguided application of Mahan's principle of decisive battle by capital ships as the beginning and end of naval strategy combined with an inferiority complex to produce a very rigid approach to naval warfare. In the upshot the strategy envisioned by the German Z-Plan—commerce raiding—was one of many strategies employed by the U.S. Navy against Japan during a long war of attrition. Japan found their critical sea communications interdicted and disrupted across their far-flung maritime empire. Japan's naval aviation, their merchant marine, and many of their island naval bases—arguably their most precious assets in a war with the United States—had been destroyed, occupied, or bypassed.

By the last year of the Pacific War, Japan's vaunted surface fleet was still largely intact—but also largely irrelevant.

The course of innovation in the three naval powers discussed was set, ultimately, by the strategic context within which it occurred. This context was for most of the period a product of the treaty system. The experience of the other naval powers illuminates innovation in the U.S. Navy in a number of ways. Above all, the U.S. experience, when compared to the others, suggests that the United States had an organizational edge over the other navies. It was the assignment of significant service responsibilities for naval aviation outside of the navy hierarchy that impeded the development of naval aviation in the British and German navies. In the American case, the establishment of the Bureau of Aviation and the decisions of the Morrow Board favored the development of naval aviation.

In comparison with the Japanese experience, the U.S. Navy's environment of organizational cooperation and team work contrasts sharply with the ideological and bureaucratic wars that ultimately proved detrimental to the Japanese. Americans have often been quick to criticize their own bureaucratic dysfunctionalism—but compared to the Japanese it was a model of organizational health. The Japanese structure, with a navy ministry where policy and resource decisions were made, and a naval general staff that planned around these decisions, had the same potential advantages as the American system. Instead, these two organizations, unlike OpNav and the General Board, were often at loggerheads with each other for much of the period. This ultimately undermined Japan's ability to plan at the strategic and operational levels, despite having a very capable and well-trained fleet.

In all of the navies the specter of the battleship loomed large throughout the period. Ironically, however, the battleship lost ground during the period that it might have regained after the ambivalent experience of World War I. In Germany, the sheer lack of time and material resources forced the Germans to a more practical and appropriate naval strategy. This in turn drove innovation. Battleships were not overemphasized and once the war began their construction was halted altogether. On the other hand, the Americans were not denied their battleships but the fortification clause severely limited how the United States could use them, at least in the opening phase of a Pacific campaign. Without logistic support, the battleship was clearly vulnerable. So, for different reasons, the Americans also developed a more practical and realistic strategy that relied on industrial production and strategic mobility. The Japanese were less flexible and more committed to the battleship in spite of their having a considerable carrier force. Their attack on the U.S. battle line at Pearl Harbor only reinforced their own incorrect assumptions and spurred the Americans toward their correct inclinations.

All four navies were constrained by the treaty system. Certainly Japan and Germany were justified in feeling that they were unfairly constrained by the Treaties. However, the fundamental logistic constraint—the fortification clause—that applied uniquely to the United States was the margin of difference that allowed the U.S. Navy to innovate and plan most realistically for World War II in comparison to the others. Sooner, in the case of the Japanese and the Germans, and later in the case of Great Britain (and the United States), naval limitation came to be regarded as the legitimate sphere for renewed competition.

Chapter 9

Conclusion

The enemy of our games was always—Japan—and the courses were so thorough that after the start of World War II—nothing that happened in the Pacific was strange or unexpected. Each student was required to plan logistic support for an advance across the Pacific—and we were well prepared for the fantastic logistic efforts required to support the operations of the war.—The need for mobile replenishment at sea was foreseen—and practiced by me in 1937. . . .

Fleet Adm. Chester W. Nimitz, 1965

Article XIX of the Washington Naval Treaty presented the U.S. Navy with a specific strategic challenge within the context of a war with Japan. The constraints posed by the fortification clause modified Navy leaders' institutional conception of sea power. The original or traditional conception required "bases far and wide" but Article XIX of the Washington Naval Treaty had drastically curtailed the Navy's ability to develop new bases or fortify existing ones in the western Pacific. The fortification clause crystallized for the Navy the strategic problem it must address in projecting and maintaining naval power in the event of war in the western Pacific. This problem—or the view of strategy and sea power that gave rise to its perception—was not shared by the civilian leadership and administrations of the period. The civilian secretaries of the Navy often had little interest in developing ways around the Treaty and its attendant fortification constraints. They, and the presidents of the period (including FDR) were more interested in how to expand and extend the naval limitation regime and how to build the treaty fleet at a minimum expense. However, despite this dichotomy in strategic perceptions, the U.S. Navy's leadership—especially the General Board—

remained nearly unanimous in believing that the fortification clause posed a vexing problem of the first magnitude for the treaty-limited Navy during the period of the treaty system.

To this end, the fortification clause played a major role in the development of strategy and in the design of the force structure for a treaty fleet. This force structure was not just about battleships and aircraft carriers. As reflected in the *1922 U.S. Naval Policy*, the plan to build a treaty fleet was complex and robust. Complexities involved with the implementation of the Treaties, and other ship programs, occupied the attention of the Navy's leadership as much as did battleships and aircraft carriers. The underlying basis for the construction of the fleet was the Washington Naval Treaty, its strategic implications, and the Navy's conception of sea power.

* * *

The interwar budgets available to build the Navy's treaty fleet were constrained during the 1920s more by ideological and political factors than by economic conditions. The treaty system's perceived potential to limit and scale back naval expenditures played a profound role in generating both legislative and executive expectations that were at odds with the institutional Navy's construction recommendations. Navy leaders clearly thought these recommendations were the minimum to adequately defend what they perceived to be U.S. interests. Ironically, this parsimonious ideological environment seemed to favor creativity and innovation.

Battleship modernization represented the Navy's first response to the Washington Naval Treaty. Given the limited scope for reconstruction allowed by the Treaty, the Navy was restricted to making incremental improvements, which had their greatest effect at the tactical and operational levels of war. At the operational level, the requirements mandated by the new strategic realities imparted by the fortification clause caused the Navy to first address the inadequate coal-propulsion systems of its six oldest dreadnoughts. Most battleship improvements, however, focused on the tactical level. They were explicitly authorized by the Treaty but also came to represent an area of controversy.

The reconstruction clause of the Washington Naval Treaty specifically allowed three thousand tons of extra weight per capital ship for antiair and antisubmarine improvements. In addition to strict self-defense improvements that reflect the tactical level of war, the modifications recommended by the General Board also affected cruising radius. It was soon discovered that internal redesign of fuel compartments could be justified under the heading of antisubmarine protection. More controversial was the Navy's liberal interpretation of the Treaty's language on turret modifications. Although initially stymied by British protests, the Navy was able to finally prevail over against stiff opposition by the secretary of the Navy in getting support for these modifications, but not for the oldest battleships.

The battleship reconstruction program continued from 1923 until war broke out with Japan, but most budgetary fights over it occurred during the early to mid-1920s. Once these programs were funded the Navy, primarily via the General Board, attention was turned to the more pressing issue of building the remainder of the treaty fleet to augment the core battleship force. The Navy's priorities for this fleet centered on naval aviation, cruisers, submarines, destroyers, and new types of logistic support ships and programs. The frustration born out of the slow delivery of aircraft carriers to the fleet resulted in some unique solutions whose purpose was to provide air cover for the fleet. The Navy recognized rather quickly the profound implication the fortification clause posed to its traditional concept of land-based maintenance support to the fleet. As early as 1923, the Navy had already developed an innovative new strategic concept built around mobile bases. This concept was reflected in war planning, doctrine, and building program recommendations by the General Board.

Like the battleship program, naval aviation's future was in many ways already established by the terms of the Washington Treaty. The two battle cruisers converted to aircraft carriers ended up being delayed in delivery. They in turn inadvertently delayed the delivery of additional carriers to the fleet because of the slow pace of their conversion. This slow pace was, for the most part, due to anemic funding for the conversions. In the meantime, the Navy tried to come up with ways to creatively base its Asiatic Fleet aviation aboard a variety of converted auxiliaries as well as advocating the construction of purpose-built airplane tenders that would service not just the Asiatic Fleet but the battle and scouting fleets. This program, too, ran into delays and compromises that frustrated fleet commanders, war planners, and the members of the General Board.

The Navy introduced new language at the London Naval Conference of 1930 that offered another means to get more naval aviation to the fleet—a fleet that lacked air bases ashore in the Far East due to the fortification clause. As already shown, Admiral Moffett of BuAer and others inserted a clause into the London Naval Treaty of 1930 that allowed 25 percent of a nation's light cruiser quota to be of the new type known as a flying deck cruiser. Although Moffett's intent was probably to disguise a carrier as a cruiser, the rest of the Navy seemed to see this new vessel as a promising means to solve a number of operational challenges associated with War Plan Orange. Although never built, the flying deck cruiser was an example of the creativity and the receptivity of the Navy to innovative ideas during a period of fiscal and material poverty. The flying deck cruiser may have been an idea that was simply ahead of its time. Once the vexing problem of aircraft carrier delivery was ameliorated under FDR, the money that might have built the flying deck vessel seemed better spent on new carrier construction. This episode shows how the so-called "battleship admirals" had shifted in their thinking about aviation. They had come to a position where they supported carrier construction levels on a par with, and even in excess of, what they wanted for battleship construction once the capital ship building holiday had expired.

It was at the strategic level of war that the Navy perhaps innovated most dynamically. The secret mobile base project (MBP) and the associated General Board hearings on the massive floating dry docks had no precedents in either naval planning or construction. The original plan for sectional floating dry docks was eventually adopted, but only after years of budgetary neglect by Congress and the administration. Nevertheless, by the early 1930s the Navy had spent precious dollars to build a small deployable floating dry dock. ARD-1 was the first floating dry dock built in almost thirty years. By the advent of World War II, the Navy had already started to build the largest mobile floating dry dock in the world—the Advanced Base Sectional Dock (ABSD-1)—some of whose components still existed in the twenty-first century. If any single program symbolizes the Navy's changed way of thinking during the treaty period, it might be these sea-going maintenance yards.

Ideological Factors Affecting Innovation

The Navy underwent an evolutionary transformation in its thinking about sea power during the interwar period. The officer corps' institutional understanding of sea power was based primarily upon the writings of A. T. Mahan. Navy officers saw sea power as being composed primarily of three elements: a fleet of warships, a merchant marine fleet and trade routes, and domestic and overseas bases. By the time of the interwar period, this concept had virtually achieved the level of an ideology—a body of ideas reflecting the institutional (and professional) needs of the Navy's officer corps.[1] It is evident that the Navy's concept of sea power was constrained by the Washington Treaty in two areas—the fleet and overseas bases. The fleet was constrained by the majority of the clauses in the Treaty—the capital ship holiday, the 5–5–3 ratio, tonnage limits, and specific limits on sizes and armaments of cruisers and aircraft carriers. The London Treaty later broadened these constraints to include the auxiliary classes not limited in the original Washington Naval Treaty. However, these treaties did not quantitatively negate the U.S. fleet's superiority vis-à-vis the Japanese. Rather they established a basis to maintain a 5–3 margin of superiority in most warship categories. The fleet element in the Navy's conception of sea power was limited but not fundamentally jeopardized.

This was not the case with bases. The United States' interests in the western Pacific under this view of sea power mandated that a number of support bases be built or improved in order to better support the fleet in an overseas campaign against the Japanese. These bases need not be built if the region could be stabilized and the Japanese accommodated for their "inferior" naval position. This argument, perhaps rightly, won the day at Washington in 1922. However, the Treaty formalized a policy of status quo—not building or improving existing bases—because the Japanese simply did not believe the United States would settle for the naval parity in the western Pacific. Status quo via a fortification clause was effectively made international law as a compromise to get the Japanese to agree to an "inferior" fleet. The fortification clause, in stark contrast to the other provisions of the Treaty, directly challenged the Navy's conception of sea power. How could the Navy succeed without adequate bases? Instead of recasting their conception of sea power, Navy leaders instead clung to it. Instead of a revolution in strategic vision, they sought an innovative transformation of the means by which to achieve their traditional vision. They planned to ameliorate the impact of the fortification clause via two means. The first means was straightforward and involved redoubled efforts to improve the efficiency of the fleet. These efforts led to a series of incremental improvements to the fleet which are best reflected by the modernization of the battleship fleet and the design of 10,000-ton cruisers. These design changes resulted in a Navy with endurances, for most of the major vessels, in excess of 10,000 nm. Similarly, the U.S. Navy developed some of the longest range destroyers (*Farragut* and *Mahan* classes) and submarines (*Gato* class) during the interwar period.[2]

The second means had to do with the Navy's traditional understanding of basing. Naval aviation developments, the MBP, and the General Order, which established the Fleet Base Force were manifestations of Navy officers' efforts to circumvent the fortification clause through creative aviation basing schemes, logistics, and organization. Shore-based solutions were no longer adequate. Now officers considered more seriously basing them at sea, "sea-basing" in today's terminology, or making bases mobile enough to follow the fleet for rapid establishment at interim or temporary locations.[3] The Navy was already proceeding along a trajectory of policies that emphasized putting as much aviation at sea as possible. However, the logistical means to support aviation afloat over the long term—due to aviation's newness and the inexperience of naval officers—was nowhere near maturity. The fortification clause channeled the Navy further down the path of primarily sea-based aviation because it denied the option of shore-basing aircraft at the beginning of a western Pacific campaign.

The Navy was only just beginning to seriously consider the logistical ramifications of an overseas naval campaign at the end World War I. The ONI study by Frost et al. and that of Major Ellis of the Marine Corps were pre-treaty attempts to conceptualize a naval war over the vast reaches of the Pacific. Here the status quo on fortifications was most influential in causing Navy leaders, especially the members of the General Board, to conceptualize support and maintenance for ships using primarily sea-based means. Extensive efforts were made to retain the old paradigm of fixed shore facilities. Orange planning retained a notion that America's "Port Arthur" in Manila Bay could be held until the fleet arrived. Not until the 1930s did the Navy institutionally concede that it was unlikely that Manila could be held and that the United States would have to adopt a methodical "step-by-step" advance across the Pacific to recapture the Philippines. In the meantime, incremental planning had generated programs that established the basis for a successful U.S. strategy in the Pacific theater in World War II. At-sea and mobile logistics were first envisioned as being set up at a "western base" at some other location in a part of the Philippines that might still be in U.S. hands when the fleet arrived. This base would be built around the mobile base project first conceptualized in 1923. When this no longer seemed likely, the "western base" became an advanced base that would be established somewhere in the Marshall or Caroline Archipelagoes, perhaps the Mortlock group south of Truk. Here, too, the mobile base served as the template for an advanced base. The fortification clause channeled, refined, and drove this strategic process. As it turned out, the United States ended up establishing the initial elements of the mobile base in 1942 in the New Hebrides near Espiritu Santo in the South Pacific.

Also important in the U.S. Navy of this period was its obsession with efficiency. The Navy of this period was not so much a cult of the battleship as a cult of efficiency. Dudley Knox, the head of OpNav's historical section, had predicted that the Treaty would only spur navies to rechannel competition by attempting to increase efficiency, especially his own Navy. In opening the Board's first extensive hearing on the implementation of the Treaty, Admiral Rodgers reminded the Board of their mandate to " . . . maintain efficiency during a period of peace, which will in all probability last for some years."[4] The Navy's quest for efficiency was buttressed by its view of itself as a problem-solving culture. The Navy's approach to war fighting problems and planning—strategic, operational, and tactical—was to apply the methodical process that was taught at the Naval War College. However this was not the only element in the process used by the Navy to solve problems. Conceptual solutions were then tested in Fleet Battle Problems or exercises. These results were studied and only then translated into practical and engineering solutions that were reflected in force structure (e.g., the *1922 U.S. Naval Policy*) and ship design. Navy officers were comfortable with these processes. The General Board hearings reflect the generally optimistic view these officers had that they could come up with the right engineering and tactical solutions for their treaty fleet and for individual ship designs—if only the civilians would pay for them. The implementation of the Washington and later treaties highlights both of these approaches. In particular, the fortification clause caused the Navy to immediately alter its strategic planning problems for the curricula at the Naval War College and in the planning by the War Plans Division. The clause also had a direct impact on the engineering and design for everything from battleships to floating dry docks.

The Navy's conception of sea power and its utility was affected by other factors as well. Among these was the Navy's angst with regard to its view of the strategic challenges posed after the Washington Conference. In contrast to their optimism with respect to design solutions, Navy leaders, with some exceptions, were pessimistic about the prospects for long-term stability and security in the Pacific. This view had its basis, often articulated in language at the beginning of each annual General Board building-policy serial, in the view that conflict and competition between nations was endemic to the human experience. Admiral William Rodgers noted in 1921 that "the world has done little to settle into peace . . . the governments of the world apparently do not yet see their way clear to yield to the insistent demands for disarmament. . . . No nation seems willing to sacrifice any actual position of equality or superiority for war which she now holds." Most Navy leaders only grudgingly entertained the notion that arms limitation might work and their spokesman on this point was again Dudley Knox. They especially wanted to accumulate more evidence that naval limitation had merit before continuing with a regime of further limitations and building "holidays."[5] They articulated a view that U.S. policies vis-à-vis China and the Philippines would require that the treaty fleet be built as quickly as possible in order to guarantee the peace, otherwise the United States would undermine its ability to protect its interests as well as the strength of its position at future arms conferences. These convictions led to a sense of purpose born out of anxiety instead of a more optimistic view that might have lessened the urgency for innovation.

Treaty-Related Factors

The treaty system does not qualify as an independent factor affecting innovation. Rather this dynamic system influenced interrelated factors involving strategic, fiscal, and materiel considerations. The interdependency of these factors is all-important to understanding their influence on innovation. Strategy relies on resources to achieve policy goals. Said another way, ends (policy goals) require ways (strategy and plans) and means (resources) for their accomplishment. Strategy only has meaning when it is animated with money and materiel. Finally, the money comes from a political process. A more subtle factor was the timing of the Washington Conference, occurring shortly after World War I, while the lessons of that conflict were still fresh in the collective memories of naval officers.

The treaty system established the strategic framework for policy. As we have seen, the most striking effect on Pacific strategy was derived from the fortification clause, which caused the Navy to significantly revise its strategic plans. Strategic revision of the war plans occurred in the OpNav War Plans Division. This division generated plans, now limited by the Treaty, which established the types and numbers of ships and a corresponding construction timetable. However, the War Plans Division did not plan in a vacuum. Although no longer officially responsible for war planning, the General Board was still the organization responsible for building the force structure called for by these plans. They determined what was recommended for construction in the budget and what the priorities were to be. Interaction between the Board and the OpNav planners occurred mostly during Board hearings. Although generated in OpNav, war plans were often revised because of the dynamic interaction with the General Board. Above the level of the secretary of the Navy, or higher, there still remained the difficult process of the authorization and appropriation (or allocation) of the funds for new construction. Here, too, the Treaty influenced what was and was not finally included in the budget.

It was very rare that the General Board ever got everything it wanted for a particular fiscal year. More often it got considerably less, which in turn influenced the feasibility of adhering to the timetables and numbers dictated by the war plans. The timelines for these plans, as reflected by the mobile base project for example, relied on peacetime construction. War plans should have been rewritten to better reflect political realities, but often they were not. Instead the General Board was forced to scale back requirements while still attempting to best meet the intent of the war plans. The Board's members were uniquely positioned to do this given their intimate understanding of the naval treaties' clauses. The War Plans Division envisioned a naval campaign in the distant reaches of the Pacific—somewhere between the eastern extremity of the Marshalls to Manila Bay. The planners did not substantively change this requirement, even after the formal adoption of the "step-by-step" strategy in the mid-1930s. The General Board, in response, felt it incumbent to try and build a treaty fleet that would enable this strategy. Because of the fortification clause, the Board had to emphasize long radius of action, strategically mobile logistics, and at-sea aviation support in its design recommendations and guidance.

In the 1920s Navy budgets depended more on the political ideology of many members of Congress and the executive branch leadership and less on the existing economic situation, which was favorable. President Harding and Secretary of State Hughes were among the architects of a system of naval limitation that they hoped would lead to a trend of decreasing naval expenditures and not just a one-time savings. Decreased naval expenditures would in turn contribute to what they saw as an environment of expanding economic opportunity and prosperity. Men like Hughes and Harding were also sincerely motivated by a desire for peace. It was no accident that they met the General Board's requests for an ambitious program to build a robust treaty fleet with both skepticism and fiscal restraint.

The executive branch's recommendations for naval construction, especially those of the Republican administrations of the period, were met with additional skepticism by Republican Congresses for much of the period. At times congressional economizing in naval expenditures exceeded that of the executive branch both in their zeal for savings and in their convictions about the security promised by universal disarmament. For Congresses and the administrations of this period, especially throughout the 1920s, every new opportunity for an arms limitation or disarmament conference was welcomed as an opportunity to realize both further savings as well as a chance to further develop a system of international disarmament and security. Only when these hopes were rebuffed, as at Geneva in 1927, did the purse strings then loosen and result in partial implementation of the General Board's recommendations for naval construction.

By the 1930s this dynamic had become more complicated by the Great Depression. Now Hoover's policy of government economy for its own sake, as a way of "setting a good example," came to play as much of a role as other ideologically derived fiscal policies. In this light the London Naval Treaty might be viewed as a way of responding to the Depression by eliminating a loophole in the original naval treaty that allowed unrestricted construction of auxiliary warships as well as a means to use expanded naval disarmament as a guarantor of security. Ironically, the impact of these increasingly constrained fiscal environments on innovation tended to be positive. Navy leaders grew frantic as they faced the prospect of budget cuts that would curtail both construction and operations. They faced the specter of dealing with an international crisis without a fleet adequate to meet the challenge. Navy leaders—for example, on the General Board and in the bureaus—became willing to entertain unfamiliar and nontraditional solutions to their problems. Flying deck cruisers, giant mobile dry docks, battleships as shore bombardment platforms, carrier task forces, and submarine war against commerce and enemy invasion forces were among the ideas that were seriously entertained by the Navy's senior leaders and planners. Eventually these new ideas became embedded in the psyche of the bulk of the Navy's officer corps.

Instead of going back to the old way of doing business the Navy had evolved a measurably different way of doing business when fiscal policy did change for the better during the FDR's administration. This new way of thinking about naval power involved a commitment to programs and doctrines that had first found merit during the days of treaty- and policy-related budgetary neglect. However, instead of "happy days are here again," the Navy continued with its original plan outlined in the *1922 U.S. Naval Policy* to build a treaty fleet for an Orange conflict. The Navy's conception of what this treaty fleet should consist of, though, had measurably changed. Battleships had declined somewhat in their importance—especially as to their employment in the early days of a Pacific campaign. Ideas and programs that might have been abandoned or less emphasized had budgets been different—for example, the mobile base plan and naval aviation—had become the centerpieces of strategy, operations, and naval construction.

In some cases, such as the flying deck cruiser, the Navy did abandon some of the more "radical" solutions that extremity had led it to investigate. However, it was during the Treaty period that the bulk of these important changes in individual, organizational, and ultimately institutional Navy attitudes took place. When the time finally came and the Navy was allowed to construct a "non-treaty" fleet, Navy leaders had a more realistic appreciation for the kind of fleet they needed to build to wage war in the Pacific. This appreciation was not the result of an instantaneous, collective realization, but rather had its roots in the experience of building a treaty fleet for an Orange strategy constrained by the treaty system.

The end of the treaty system may have had a negative effect on the U.S. Navy's innovative efforts after an era of fiscal, strategic, and political constraints. Many of the solutions to the problems of the 1920s and early 1930s remained in place after 1937. This was because the Treaty-inspired constraints did not disappear overnight. Political constraints remained in place regarding building up U.S. bases in Guam and the Philippines. Political factors in turn tended to reinforce the strategic constraint attributable to the fortification clause that still made at-sea basing attractive. The worsening world situation in the 1930s and the passing of the treaty system eased fiscal constraints. However, increased funding went toward gaining lost ground rather than on the initiation of completely new programs. When the money was available, for example, to build flying deck cruisers (which Admiral King still supported in 1940), it instead went toward building conventional cruisers. Only the war would cause the Navy to convert these hulls into light aircraft carriers.

Another way in which naval treaties, especially the Washington Treaty, might have influenced innovation in a positive way has to do with the question of "lessons learned." Williamson Murray states that: "One of the most frequently quoted axioms of historians is that generals prepare for the last war and that is why military organizations have a difficult time in the next conflict. In fact, most armies do nothing of the kind, and because they have not distilled the lessons of the last war, they end up repeating most of the same mistakes."[6]

Substituting admirals for generals and keeping this in mind, the Naval Treaties might have had the unintended consequence of "freezing" some of the naval insights of World War I in place. Here the comparative aspect with the experience of the other naval powers is useful. World War I had drastically changed the face of naval warfare. By the end of that war submarines, destroyers, and—for Great Britain—aircraft carriers were established components of the world's foremost fleet, the Royal Navy. All the other naval powers, including the United States, measured themselves against this fleet. Japan and the United States had both rapidly created their own experimental aircraft carriers, *Hosho* and *Langley*.[7] Submarines had also proven their value and their terrifying potential to alter naval equations of power. They could sink battleships and in unscrupulous hands could be used for what was regarded as an illegal form of warfare against a nation's commerce. The submarine had also prompted the construction of innumerable convoy and warship escort vessels, mostly destroyers. Large inventories of these escort vessels were afloat at the end of the war.

The performance of the battleship during World War I had altered conceptions of its efficacy as well. During World War I the United States Navy was very much focused on submarine warfare. In this type of warfare the battleship was de-emphasized and even irrelevant. The United States was careful during that war to protect its rather large investment in dreadnoughts by limiting their operational use while at the same time focusing its creative energies on solving the maritime problem at hand—protection of the sea lines of communications. The solutions that the U.S. Navy tested revolved around naval aviation, destroyers, and convoy tactics with the submarine as the primary adversary. Also, due to the Anglo-German naval arms race that preceded the war, many political leaders and civilians saw the battleship as a dangerously destabilizing weapon. Other less critical observers saw it as an expensive, but indecisive, weapon. As for battle cruisers, the United States had practically abandoned this class altogether and had already started planning for the conversion of existing hulls into aircraft carriers. The Washington Conference only reinforced this trend. Limiting capital ships in Washington effectively emphasized the reduced importance of these ships as components of sea power while at the same time highlighting that they still remained the centerpiece of modern fleets.

The Washington Treaty was signed soon enough after World War I so that it effectively "froze" a force structure in place that was existent at the end of the war. Although the size of navies' capital ship forces decreased, the power of the "battle lines" relative to each other remained about the same. With capital ship tonnage frozen, and construction of new ones prohibited, the remaining fleet became the only part that naval leaders could alter either by designing and building new classes that were not limited by treaty, building up to limits for aircraft carriers, or scrapping old ships in favor of other classes. Most of the navies of the period chose to retain much of their post-war force structure—it was cheaper this way. Much of what was retained was a larger component relative to the now-reduced battleship components. For example, navy leaders may not have built the numbers of destroyers that would be needed to battle submarines for the next war, but they retained more destroyers, and probably built—especially in the U.S. case—more destroyers, than would otherwise have been the case had new battleship construction not been eliminated. The same logic holds true for cruisers.

Naval aviation also gained an acceptance among naval officers that it might otherwise not have enjoyed had it not been for the Washington and London Treaties. World War I taught navy officers that naval aviation had exceptional promise as a component of naval warfare and this lesson was reinforced by the clauses of the naval treaties. Both treaties specifically identified and limited the aircraft carrier, which tended to highlight its importance to those officers of the nations most affected by the treaties' constraints—Japan, the United States, and Great Britain. In particular, the fortification clause made it more likely, not less likely, that aviation would be embarked on ships in order to employ air power in the first phases of a western Pacific conflict.

If World War I taught that naval aviation, submarines, and destroyers were very important components in the modern naval environment, then the treaties went a long way toward codifying this view. Just as the prohibitions of Versailles focused the Germans on developing submarines, mechanization, and air power, so, too, did the naval treaties focus naval officers on other classes of ships. Reducing the numbers of battleships and then prohibiting their construction affected the interwar navies in other ways. These navies could no longer build fleets that were overwhelmingly composed of battleships. Although navies were still battleship-centric, what the treaties did do was to cause navies to focus their construction on cruisers, carriers, destroyers, submarines, and new auxiliary components such as aircraft tenders and mobile dry docks. Because of the uncertainty of the future, and especially the uncertainty associated with war, the more balanced navies were better prepared for an uncertain future. This was particularly true of the United States.

The evidence presented here suggests that treaty system played a major role in causing the United States to not only build a robust fleet, but to develop detailed plans for a power projection fleet. Power projection in this sense was less the primary offensive mechanism but rather a means to an end—a decisive fleet action. However, in developing the plans and designs to project power in order to support offensive fleet action the Navy, perhaps inadvertently, laid the groundwork for the supporting elements to assume primacy. The Navy developed plans to project power in order to seize the bases that would support offensive fleet action—a very traditional objective would be achieved by less than traditional means. In the same way, carrier aviation was developed as a supporting element and not with the idea (except for perhaps Moffett and a few others) that it would be the primary arm of naval decision. Embarked naval aviation was needed in the absence of land-based air power. In the upshot these secondary means—seizing and establishing advanced bases and carrier aviation—in wartime became the primary means with the battle fleet transforming from the arm of decision to the supporting roles of gunfire support and anti-aircraft defense we find it playing in the mature central Pacific offensives of late 1943 and 1944. In this sense, the U.S. Navy's design was consistent with Mahan's fundamental idea that "Control of the sea [was] the leading object in naval war . . . " in order to execute and protect national policy.[8]

Nevertheless, more specifically, the General Board became less focused on battleships and more focused on "auxiliaries." Yet the battleship had not been completely displaced by new affections. The overwhelming majority of naval officers continued to strenuously emphasize the battleship as the measure of national naval power. But aside from this required statement of faith, the Navy, and especially the General Board, spent the majority of their time and efforts on the balance—the "auxiliary" carriers, airplanes, cruisers, destroyers, submarines, and even the lowly floating dry dock. These were the tools they could be creative with once battleship modernization was effectively enacted and funded. Time was also a factor in the acceptance of new concepts. The more time that naval officers devoted to building, experimenting, and planning around new designs and concepts, the more committed to these same things they tended to become.

As stressed earlier, World War I was critical to the process of attitude change. World War I's influence helps to explain the partial transference in the "affections" of the naval leadership of the United States from the battleships to other platforms during the interwar period. This transference bloomed into open passion by the larger Navy after December 7, 1941. After World War I, the Navy—in social science terms—"reverted to type," meaning it returned to its prewar dreadnought, second-to-none-navy focus. "Second-to-none" clearly refers to the quality and number of dreadnought battleships. However, a return to the battleship paradigm of the prewar era was opposed by public opinion in favor of scaling back the 1916 naval construction, and, as has been shown, was a primary impetus for the Washington Conference in the first place.

The timing of the Washington Conference was fortuitous for the U.S. Navy. In November 1921 many still remembered the apparent impotency of the battleship in the late war—especially Navy officers like Pratt, Sims, Fiske, and Moffett. Thus, it was a smaller conceptual leap for the Navy in the immediate post-war period to refocus, after the disappointment of Washington, back to those bothersome auxiliaries like the airplane (and the aircraft carrier), destroyers, dirigibles, and submarines that had proved so useful in the late war. However, the impact was more profound than that. The fortification clause focused these officers on new roles for these platforms. Naval leaders and innovators, like Admiral Sims, who had so recently been focused on the protection of sea lines of communication during World War I now designed ships from the vantage point of a naval campaign along extended sea lines of communication and the projection of naval power in an austere logistics environment. Likewise, Orange War planning was adapted to the realities of a western Pacific campaign without large forward bases.

The naval arms limitation system of the interwar period ultimately had the opposite of its intended effect—instead of promoting peace it channeled and fostered competition that in turn fueled innovation. Nations of the interwar period tended to compete despite the best efforts of the diplomats.[9]

Other Factors

The U.S. Navy of the interwar period was predisposed toward technological solutions to the problems posed at tactical, operational, and strategic levels of war—more so than the bulk of the U.S. Army (with the significant exception of its Air Corps). This is because navies generally are technologically oriented institutions. Even more than armies, navies rely on their equipment, not only in battle but in peace, to protect them from the enemy and the sea. The development of aircraft carriers, submarines, aircraft engines, improved fire control, radio, and ever-more efficient fuel oil propulsion systems with electric drive demonstrate that the U.S. Navy routinely used technology to solve problems. In this sense the U.S. Navy was not significantly different than the other major navies of the period. However, the interwar Navy relied heavily on technology to ameliorate the fortification clause. The British and the Japanese used technology to address different problems. In the Japanese case, technology was combined with relentless training to ameliorate their "inferior" position conferred by the treaty system. In the British case, technology was used to maintain what was thought to be a superior position vis-à-vis any potential naval adversaries (except perhaps the United States).

However, technology was simply one among many factors. The Navy was already organized in a fashion that encouraged operational innovation. It had a relatively flat organizational hierarchy and a tradition of frank communication between its constituent organizations. Managerial control was spread out, during the interwar period, among a number of organizations. The sometimes depressing effect of strong executive leadership on new ideas from the "rank and file" was ameliorated to some degree by the healthy collaboration, and sometimes competition, between the General Board and OpNav. In this hierarchy the bureaus often worked for both OpNav and the General Board. Savvy leaders like Admiral Moffett would sometimes play one against the other—often with positive results. Even Admiral Pratt's reorganization of the Board was not intended so much to give additional power to OpNav as it was to ensure that the General Board's advice was not unduly influenced by a powerful CNO.

In all of this, the Naval War College maintained its organizational utility, working closely with both the General Board and OpNav on solutions to shared challenges and problems like the fortification clause. Like the General Board, the president of the College forwarded strong opinions about the efficacy of new designs such as the flight deck cruiser, doctrines, treaty initiatives, and even recommended budgetary language for the building policy. The War College often worked closely with OpNav's planners by incorporating the latest iterations of War Plan Orange in its gaming curriculum, proposing estimates of the situation, and forwarding game results to OpNav and the General Board. In turn, OpNav folded these inputs into the design of the annual Battle Problem exercises for the fleet in "the real world"—the world where ships broke down, communications failed, and the enemy did not do as he was scripted. This system is still the one used by the Navy today except that the role of the General Board has to some degree been assumed by OpNav and the Joint Staff. The efficacy of this process was proven during World War II and the Navy has seen little reason to alter it since.[10]

On balance, organizational factors played a positive role in influencing innovation during the Treaty period in the U.S. Navy. The factors of most importance were the interrelationships among Navy organizations, how members were assigned to these organizations, and the internal structures of the organizations themselves. The General Board played a central role as an organizational entity that facilitated innovation because it was here that the implementation of treaty policy intersected with force and ship design. This intersection was further positively influenced by the composition and structure of the Board. Members brought their unique experiences and relationships to the Board and continued to refer back to the Board as they went to other billets. Often their next job or their previous job involved a close association with the General Board, for example Admiral Pratt's assignment as president of the Naval War College after his stint on the General Board in the 1920s or Captain Schofield's similar assignment from the Board to Chief of Naval Operations' chief of war plans. This "job-shuffling" favored, rather than discouraged, organizational collaboration in the interwar Navy. Also, most of the officers assigned to General Board had already "bought into" the Navy's experimental approach to problem-solving that was taught and best exemplified by the War College curricula and practiced during Fleet Battle Problems.

Here again comparison with the other major navies of the period is useful. The more decentralized organizational hierarchy within the U.S. Navy seems to have served it extremely well when contrasted with the British and Japanese navies. With respect to the Japanese, the critical organizational relationship was that between the naval ministry and the naval general staff. As Sadao Asada and others have shown, this relationship was strained at the beginning of the period and only got worse as time went on. These organizational conflicts tended to dampen rather than encourage innovative thinking—especially at the higher levels of war. By 1931, after the London Treaty, the naval ministry, which had been the force for moderation in the IJN, had been effectively subsumed by the more militant members of the "fleet faction" who ran the naval general staff. The cordial and collaborative relationships between OpNav, the General Board and other naval organizations stand in sharp contrast to the Japanese case.[11]

Comparison with the British case yields a number of insights as well. Unlike the Japanese, the British organizational hierarchy was less factional than it was one-dimensional. The British Admiralty combined all of the elements of the U.S. Navy department staff, the General Board, the bureaus, and OpNav into one unitary organization. Especially stifling to innovation in naval aviation was the lack of an entity in the Royal Navy with the organizational power and independence of Moffett's BuAer. British naval aviation suffered due to its divided administrative and operational chains of command and the pernicious influence of subordinating its requirements under the air ministry, an organization completely separate in nature and function from the Royal Navy and the admiralty.[12]

In all human endeavors decisions have unintended consequences. The decisions made by the U.S. Navy and government in response to and as a result of naval arms limitation were no different. One consequence of the treaty system was to heighten the importance of the U.S. Navy in the formulation of U.S. foreign policy. The mechanism by which this occurred was possibly the reverse of the old cliché about "out of sight out of mind." Because of the exceptional diplomatic focus on arms limitation in the interwar period, the U.S. Navy was constantly "in sight and in mind." The Navy could not be ignored, as the U.S. Army often was. If land or air armaments limitation had been more successful, then the Army might have been more in the public and political eye as the Navy was. Also, the Navy later came to hold a more prominent place in the domestic policy counsels of U.S. administrations given the confluence of the Navy with nonmilitary agencies due to the National Industrial Recovery Act. In this manner, the Navy came to hold a formal place in the public and political spheres that it had not held prior to the treaties.

* * *

In 1922, the U.S. Navy conceived and implemented a design for a completely new fleet based on the Washington Treaty. With the collapse of the treaty system in 1937 the U.S. Navy finished building its treaty fleet instead of designing a completely new fleet. More aircraft carriers were built, ship-building remained concentrated on long cruising radius, *Langley* was converted to a seaplane tender and sent to the Philippines, battleship tactics remained focused on the 27,000-yard gun battle, War Plan Orange—despite being eventually replaced by the Rainbow Plans—was still the strategic template, and submarines were still officially fleet support warships whose nominal missions did not include unrestricted submarine warfare. Only in 1940, with war threatening, did the U.S. Navy apply itself to the task of building a measurably different fleet. However, this fleet was in many ways simply a larger version of the fleet it had just built and so was only an evolutionary conceptual step.

The U.S. Navy did lose some momentum in its feverish attempts to overcome what it saw as the inadequacy of the fleet after the treaty system had effectively been abandoned. With the approval of plans for the rapid expansion and modernization of the fleet in 1940, there was just too much to do operationally and in training on a day-to-day basis. However, the fleet the Japanese found moored at Pearl Harbor and in the Philippines on December 7–8, 1941 was primarily a treaty-built fleet. The long-term solutions that the Navy employed during World War II were crafted during the lean years of the treaty system and provided the template for that first Navy that fought the Japanese "to a standstill" until the "Two-Ocean" fleet came online in mid-1943. This second Navy, the most powerful blue-water navy in human history, was in part a direct result of the Navy's reaction to the fortification clause of the Treaty.[13]

This outcome was not the result of having the precise Navy needed to win World War II with the opening of hostilities, but because *the kind of Navy needed* had already been discussed and thought about extensively during the hearings of the General Board, in the classrooms at the Naval War College, at sea, and in the planning cells of OpNav's War Plans Division. The concepts and designs that emerged from the building ways after 1941 were neither new nor strange—they simply required an occasion to accelerate their design maturity and delivery. This occasion arrived in 1940, with war imminent. The operational and technological tools were already familiar, although many unknowns remained—but these were the sorts of unknowns that war always brings. Applying existing strategic, operational, and tactical solutions and then adjusting them to the realities of war came easier to Navy officers because of their focus over two decades on precisely the strategy and materiel requirements that a Pacific War without preexisting bases demanded. U.S. Navy officers were more flexible in their attitudes and their willingness to learn as a result. Part of this was undoubtedly due to the experience they gained in building a fleet constrained by the treaty system.

The Navy did not make a radical change to its paradigm of sea power during the interwar period. Instead it tried to adjust its solutions and designs to fit an existing paradigm or conception of sea power that was rooted in the teachings of A. T. Mahan. In doing this the Navy changed the boundaries, but not the essentials, of the paradigm. Fleets, maritime commerce, and basing were all still regarded as essential to the sea power equation. However, the definition of just what "basing" really meant had been expanded to include mobile bases. Both fixed and mobile bases were needed in the strategic environment of the Pacific. Although the paradigm did not change, the fleet did. This change was accelerated by the anomaly caused by the fortification clause to the Navy's traditional conception of sea power. It became axiomatic that in wartime mobility, not just of tactical and operational units but for strategic logistical capabilities, was vital. The fortification clause forced the U.S. Navy to learn how to become a global navy. In the twenty-first century this is reflected in the U.S. Navy's "sea-basing" concept.[14]

In closing, a general observation about the relationship between naval arms limitation treaties and innovation is in order. Innovation can occur in the face of constraints. When military leaders perceive that they have to solve specific problems they rarely concede defeat in the face of constraints. The German, American, and to a lesser degree Japanese experiences in the period of 1919 to 1937 show how naval innovation was affected in a positive way by physical, strategic, and materiel constraints. Their curiosity was piqued, rather than dulled, by limitation or disarmament. In the American case, new ways of applying sea power were conceptualized that might otherwise have remained unexamined. The Navy's institutional view of sea power, and thus its ideology, had changed decisively.

Appendix 1
The Washington Naval Treaty*

Treaty Between the United States of America, the British Empire, France, Italy, and Japan, Signed at Washington, February 6, 1922.

The United States of America, the British Empire, France, Italy and Japan:

Desiring to contribute to the maintenance of the general peace, and to reduce the burdens of competition in armament;

In English and French; French text not printed. Ratification advised by the Senate, Mar. 29, 1922; ratified by the President, June 9, 1923; ratifications deposited with the Government of the United States, Aug. 17, 1923; proclaimed, Aug. 21, 1923.

Have resolved, with a view to accomplishing these purposes, to conclude a treaty to limit their respective naval armament, and to that end have appointed as their Plenipotentiaries;

The President of the United States of America:
  Charles Evans Hughes,
  Henry Cabot Lodge,
  Oscar W. Underwood,
  Elihu Root,
    citizens of the United States;

---

*Note: *Papers Relating to the Foreign Relations of the United States: 1922*, Vol. 1, 247-266. Treaty Series No. 671. Accessed at http://www.ibiblio.org/pha/pre-war/1922/nav_lim.html June 10, 2008.

His Majesty the King of the United Kingdom of Great Britain and Ireland and of the British Dominions beyond the Seas, Emperor of India:
> The Right Honourable Arthur James Balfour, O. M., M. P., Lord President of His Privy Council;
> The Right Honourable Baron Lee of Fareham, G. B. E., K. C. B., First Lord of His Admiralty;
> The Right Honourable Sir Auckland Campbell Geddes, K. C. B., His Ambassador Extraordinary;

and

for the Dominion of Canada:
> The Right Honourable Sir Robert Laird Borden, G. C. M. G., K. C.;

for the Commonwealth of Australia:
> Senator the Right Honourable George Foster Pearce, Minister for Home and Territories;

for the Dominion of New Zealand:
> The Honourable Sir John William Salmond, K. C., Judge of the Supreme Court of New Zealand;

for the Union of South Africa:
> The Right Honourable Arthur James Balfour, O. M., M. P.;

for India:
> The Right Honourable Valingman Sankaranarayana Srinivasa Sastri, Member of the Indian Council of State;

The President of the French Republic:
> Mr. Albert Sarraut, Deputy, Minister of the Colonies;
> Mr. Jules J. Jusserand, Ambassador Extraordinary and Plenipotentiary to the United States of America, Grand Cross of the National Order of the Legion of Honour;

His Majesty the King of Italy:
> The Honourable Carlo Schanzer, Senator of the Kingdom;
> The Honourable Vittorio Rolandi Ricci, Senator of the Kingdom, His Ambassador Extraordinary and Plenipotentiary at Washington;
> The Honourable Luigi Albertini, Senator of the Kingdom;

His Majesty the Emperor of Japan:
  Baron Tomosaburo Kato, Minister for the Navy, Junii, a member of the
    First Class of the Imperial Order of the Grand Cordon of the
    Rising Sun with the Paulownia Flower;
  Baron Kijuro Shidehara, His Ambassador Extraordinary and
    Plenipotentiary at Washington, Joshii, a member of the First Class
    of the Imperial Order of the Rising Sun;
  Mr. Masanao Hanihara, Vice Minister for Foreign Affairs, Jushii, a
    member of the Second Class of the Imperial Order of the Rising Sun;

Who, having communicated to each other their respective full powers, found to be in good and due form, have agreed as follows:

CHAPTER I. GENERAL PROVISIONS RELATING TO THE LIMITATION OF NAVAL ARMAMENT

Article I

The Contracting Powers agree to limit their respective naval armament as provided in the present Treaty.

Article II

The Contracting Powers may retain respectively the capital ships which are specified in Chapter II, Part 1. On the coming into force of the present Treaty, but subject to the following provisions of this Article, all other capital ships, built or building, of the United States, the British Empire and Japan shall be disposed of as prescribed in Chapter II, Part 2.

In addition to the capital ships specified in Chapter II, Part 1, the United States may complete and retain two ships of the *West Virginia* class now under construction. On the completion of these two ships, the *North Dakota* and *Delaware*, shall be disposed of as prescribed in Chapter II, Part 2.

The British Empire may, in accordance with the replacement table in Chapter II, Part 3, construct two new capital ships not exceeding 35,000 tons (35,560 metric tons) standard displacement each. On the completion of the said two ships the *Thunderer*, *King George V*, *Ajax* and *Centurion* shall be disposed of as prescribed in Chapter II, Part 2.

Article III

Subject to the provisions of Article II, the Contracting Powers shall abandon their respective capital ship building programs, and no new capital ships shall be constructed or acquired by any of the Contracting Powers except replacement tonnage which may be constructed or acquired as specified in Chapter II, Part 3.

Ships which are replaced in accordance with Chapter II, Part 3, shall be disposed of as prescribed in Part 2 of that Chapter.

Article IV

The total capital ship replacement tonnage of each of the Contracting Powers shall not exceed in standard displacement, for the United States 525,000 tons (533,400 metric tons); for the British Empire 525,000 tons (533,400 metric tons); for France 175,000 tons (177,800 metric tons); for Italy 175,000 tons (177,800 metric tons); for Japan 315,000 tons (320,040 metric tons).

Article V

No capital ship exceeding 35,000 tons (35,560 metric tons) standard displacement shall be acquired by, or constructed by, for, or within the jurisdiction of, any of the Contracting Powers.

Article VI

No capital ship of any of the Contracting Powers shall carry a gun with a calibre in excess of 16 inches (406 millimetres).

Article VII

The total tonnage for aircraft carriers of each of the Contracting Powers shall not exceed in standard displacement, for the United States 135,000 tons (137,160 metric tons); for the British Empire 135,000 tons (137,160 metric tons); for France 60,000 tons (60,960 metric tons); for Italy 60,000 tons (60,960 metric tons); for Japan 81,000 tons (82,296 metric tons).

Article VIII

The replacement of aircraft carriers shall be effected only as prescribed in Chapter II, Part 3, provided, however, that all aircraft carrier tonnage in existence or building on November 12, 1921, shall be considered experimental, and may be replaced, within the total tonnage limit prescribed in Article VII, without regard to its age.

Article IX

No aircraft carrier exceeding 27,000 tons (27,432 metric tons) standard displacement shall be acquired by, or constructed by, for or within the jurisdiction of, any of the Contracting Powers.

However, any of the Contracting Powers may, provided that its total tonnage allowance of aircraft carriers is not thereby exceeded, build not more than two aircraft carriers, each of a tonnage of not more than 33,000 tons (33,528 metric tons) standard displacement, and in order to effect economy any of the Contracting Powers may use for this purpose any two of their ships, whether constructed or in course of construction, which would otherwise be scrapped under the provisions of Article II. The armament of any aircraft carriers exceeding 27,000 tons (27,432 metric tons) standard displacement shall be in accordance with the requirements of Article X, except that the total number of guns to be carried in case any of such guns be of a calibre exceeding 6 inches (152 millimetres), except anti-aircraft guns and guns not exceeding 5 inches (127 millimetres), shall not exceed eight.

Article X

No aircraft carrier of any of the Contracting Powers shall carry a gun with a calibre in excess of 8 inches (203 millimetres). Without prejudice to the provisions of Article IX, if the armament carried includes guns exceeding 6 inches (152 millimetres) in calibre the total number of guns carried, except anti-aircraft guns and guns not exceeding 5 inches (127 millimetres), shall not exceed ten. If alternatively the armament contains no guns exceeding 6 inches (152 millimetres) in calibre, the number of guns is not limited. In either case the number of anti-aircraft guns and of guns not exceeding 5 inches (127 millimetres) is not limited.

Article XI

No vessel of war exceeding 10,000 tons (10,160 metric tons) standard displacement, other than a capital ship or aircraft carrier, shall be acquired by, or constructed by, for, or within the jurisdiction of, any of the Contracting Powers. Vessels not specifically built as fighting ships nor taken in time of peace under government control for fighting purposes, which are employed on fleet duties or as troop transports or in some other way for the purpose of assisting in the prosecution of hostilities otherwise than as fighting ships, shall not be within the limitations of this Article.

Article XII

No vessel of war of any of the Contracting Powers, hereafter laid down, other than a capital ship, shall carry a gun with a calibre in excess of 8 inches (203 millimetres).

Article XIII

Except as provided in Article IX, no ship designated in the present Treaty to be scrapped may be reconverted into a vessel of war.

Article XIV

No preparations shall be made in merchant ships in time of peace for the installation of warlike armaments for the purpose of converting such ships into vessels of war, other than the necessary stiffening of decks for the mounting of guns not exceeding 6 inch (152 millimetres) calibre.

Article XV

No vessel of war constructed within the jurisdiction of any of the Contracting Powers for a non-Contracting Power shall exceed the limitations as to displacement and armament prescribed by the present Treaty for vessels of a similar type which may be constructed by or for any of the Contracting Powers; provided, however, that the displacement for aircraft carriers constructed for a non-Contracting Power shall in no case exceed 27,000 tons (27,432 metric tons) standard displacement.

Article XVI

If the construction of any vessel of war for a non-Contracting Power is undertaken within the jurisdiction of any of the Contracting Powers, such Power shall promptly inform the other Contracting Powers of the date of the signing of the contract and the date on which the keel of the ship is laid; and shall also communicate to them the particulars relating to the ship prescribed in Chapter II, Part 3, Section I (b), (4) and (5).

Article XVII

In the event of a Contracting Power being engaged in war, such Power shall not use as a vessel of war any vessel of war which may be under construction within its jurisdiction for any other Power, or which may have been constructed within its jurisdiction for another Power and not delivered.

Article XVIII

Each of the Contracting Powers undertakes not to dispose by gift, sale or any mode of transfer of any vessel of war in such a manner that such vessel may become a vessel of war in the Navy of any foreign Power.

Article XIX

The United States, the British Empire and Japan agree that the status quo at the time of the signing of the present Treaty, with regard to fortifications and naval bases, shall be maintained in their respective territories and possessions specified hereunder:

1. The insular possessions which the United States now holds or may hereafter acquire in the Pacific Ocean, except a. those adjacent to the coast of the United States, Alaska and the Panama Canal Zone, not including the Aleutian Islands, and b. the Hawaiian Islands;
2. Hong Kong and the insular possessions which the British Empire now holds or may hereafter acquire in the Pacific Ocean, east of the meridian of 110° east longitude, except a. those adjacent to the coast of Canada, b. the Commonwealth of Australia and its Territories, and c. New Zealand;
3. The following insular territories and possessions of Japan in the Pacific Ocean, to wit: the Kurile Islands, the Bonin Islands, Amami-Oshima, the Loochoo Islands, Formosa and the Pescadores, and any insular territories or possessions in the Pacific Ocean which Japan may hereafter acquire.

The maintenance of the status quo under the foregoing provisions implies that no new fortifications or naval bases shall be established in the territories and possessions specified; that no measures shall be taken to increase the existing naval facilities for the repair and maintenance of naval forces, and that no increase shall be made in the coast defences of the territories and possessions above specified. This restriction, however, does not preclude such repair and replacement of worn-out weapons and equipment as is customary in naval and military establishments in time of peace.

Article XX

The rules for determining tonnage displacement prescribed in Chapter II, Part 4, shall apply to the ships of each of the Contracting Powers.

CHAPTER II. RULES RELATING TO THE EXECUTION OF THE TREATY-DEFINITION OF TERMS

PART 1. Capital Ships Which May be Retained by the Contracting Powers

In accordance with Article II ships may be retained by each of the Contracting Powers as specified in this Part.

SHIPS WHICH MAY BE RETAINED BY THE UNITED STATES

Name:	Tonnage:
Maryland	32,600
California	32,300
Tennessee	32,300
Idaho	32,000
New Mexico	32,000
Mississippi	32,000
Arizona	31,400
Pennsylvania	31,400
Oklahoma	27,500
Nevada	27,500
New York	27,000
Texas	27,000
Arkansas	26,000
Wyoming	26,000
Florida	21,825
Utah	21,825
North Dakota	20,000
Delaware	20,000
Total Tonnage	500,650

On the completion of the two ships of the *West Virginia* class and the scrapping of the *North Dakota* and *Delaware*, as provided in Article II, the total tonnage to be retained by the United States will be 525,850 tons.

SHIPS WHICH MAY BE RETAINED BY THE BRITISH EMPIRE

Name:	Tonnage:
Royal Sovereign	25,750
Royal Oak	25,750
Revenge	25,750

Resolution	25,750
Ramilies	25,750
Malaya	27,500
Valiant	27,500
Barham	27,500
Queen Elizabeth	27,500
Warsprite	27,500
Benbow	25,000
Emperor of India	25,000
Iron Duke	25,000
Marlborough	25,000
Hood	41,200
Renown	26,500
Repulse	26,500
Tiger	28,500
Thunderer	22,500
King George V	23,000
Ajax	23,000
Centurion	23,000
Total Tonnage	580,450

On the completion of the two new ships to be constructed and the scrapping of the *Thunderer*, *King George V*, *Ajax* and *Centurion*, as provided in Article II, the total tonnage to be retained by the British Empire will be 558,950 tons.

SHIPS WHICH MAY BE RETAINED BY FRANCE

Name:	Tonnage: (metric tons)
Bretagne	23,500
Lorraine	23,500
Provence	23,500
Paris	23,500
France	23,500
Jean Bart	23,500
Courbet	23,500
Condorect	18,900
Diderot	18,900
Voltaire	18,900
Total Tonnage	221,170

France may lay down new tonnage in the years 1927, 1929, and 1931, as provided in Part 3, Section II.

SHIPS WHICH MAY BE RETAINED BY ITALY

Name:	Tonnage: (metric tons)
Andrea Doria	22,700
Caio Duilio	22,700
Conte Di Cavour	22,500
Giulio Cesare	22,500
Leonardo Da Vinci	22,500
Dante Alighieri	19,500
Roma	12,600
Napoli	12,600
Vittorio Emanuele	12,600
Regina Elena	12,600
Total tonnage	182,800

Italy may lay down new tonnage in the years 1927, 1929, and 1931, as provided in Part 3, Section II.

SHIPS WHICH MAY BE RETAINED BY JAPAN

Name:	Tonnage: (metric tons)
Mutsu	33,800
Nagato	33,800
Hiuga	31,260
Ise	31,260
Yamashiro	30,600
Fu-So	30,600
Kirishima	27,500
Haruna	27,500
Hiyei	27,500
Kongo	27,500
Total tonnage	301,320

PART 2. Rules for Scrapping Vessels of War

The following rules shall be observed for the scrapping of vessels of war which are to be disposed of in accordance with Articles II and III.

I. A vessel to be scrapped must be placed in such condition that it cannot be put to combatant use.
II. This result must be finally effected in any one of the following ways:
   (a) Permanent sinking of the vessel;
   (b) Breaking the vessel up. This shall always involve the destruction or removal of all machinery, boilers and armour, and all deck, side and bottom plating;
   (c) Converting the vessel to target use exclusively. In such case all the provisions of paragraph III of this Part, except sub-paragraph (6), in so far as may be necessary to enable the ship to be used as a mobile target, and except sub-paragraph (7), must be previously complied with. Not more than one capital ship may be retained for this purpose at one time by any of the Contracting Powers.
   (d) Of the capital ships which would otherwise be scrapped under the present Treaty in or after the year 1931, France and Italy may each retain two seagoing vessels for training purposes exclusively, that is, as gunnery or torpedo schools. The two vessels retained by France shall be of the *Jean Bart* class, and of those retained by Italy one shall be the *Dante Alighieri*, the other of the *Giulio Cesare* class. On retaining these ships for the purpose above stated, France and Italy respectively undertake to remove and destroy their conning-towers, and not to use the said ships as vessels of war.
III. (a) Subject to the special exceptions contained in Article IX, when a vessel is due for scrapping, the first stage of scrapping, which consists in rendering a ship incapable of further warlike service, shall be immediately undertaken.
   (b) A vessel shall be considered incapable of further warlike service when there shall have been removed and landed, or else destroyed in the ship:
      (1) All guns and essential portions of guns, fire-control tops and revolving parts of all barbettes and turrets;
      (2) All machinery for working hydraulic or electric mountings;
      (3) All fire-control instruments and range-finders;
      (4) All ammunition, explosives and mines;

(5) All torpedoes, warheads and torpedo tubes;
(6) All wireless telegraphy installations;
(7) The conning tower and all side armour, or alternatively all main propelling machinery; and (8) All landing and flying-off platforms and all other aviation accessories.

IV. The periods in which scrapping of vessels is to be effected are as follows:
   (a) In the case of vessels to be scrapped under the first paragraph of Article II, the work of rendering the vessels incapable of further warlike service, in accordance with paragraph III of this Part, shall be completed within six months from the coming into force of the present Treaty, and the scrapping shall be finally effected within eighteen months from such coming into force.
   (b) In the case of vessels to be scrapped under the second and third paragraphs of Article II, or under Article III, the work of rendering the vessel incapable of further warlike service in accordance with paragraph III of this Part shall be commenced not later than the date of completion of its successor, and shall be finished within six months from the date of such completion. The vessel shall be finally scrapped, in accordance with paragraph II of this Part, within eighteen months from the date of completion of its successor. If, however, the completion of the new vessel be delayed, then the work of rendering the old vessel incapable of further war-like service in accordance with paragraph III of this Part shall be commenced within four years from the laying of the keel of the new vessel, and shall be finished within six months from the date on which such work was commenced, and the old vessel shall be finally scrapped in accordance with paragraph II of this Part within eighteen months from the date when the work of rendering it incapable of further warlike service was commenced.

PART 3. Replacement

The replacement of capital ships and aircraft carriers shall take place according to the rules in Section I and the tables in Section II of this Part.

SECTION I. RULES FOR REPLACEMENT
   (a) Capital ships and aircraft carriers twenty years after the date of their completion may, except as otherwise provided in Article VIII and in the tables in Section II of this Part, be replaced by new construction, but within the limits prescribed in Article IV and Article VII. The keels of such new construction may, except as otherwise provided in Article VIII

and in the tables in Section II of this Part, be laid down not earlier than seventeen years from the date of completion of the tonnage to be replaced, provided, however, that no capital ship tonnage, with the exception of the ships referred to in the third paragraph of Article II, and the replacement tonnage specifically mentioned in Section II of this Part, shall be laid down until ten years from November 12, 1921.

(b) Each of the Contracting Powers shall communicate promptly to each of the other Contracting Powers the following information:
  (1) The names of the capital ships and aircraft carriers to be replaced by new construction;
  (2) The date of governmental authorization of replacement tonnage;
  (3) The date of laying the keels of replacement tonnage;
  (4) The standard displacement in tons and metric tons of each new ship to be laid down, and the principal dimensions, namely, length at waterline, extreme beam at or below waterline, mean draft at standard displacement;
  (5) The date of completion of each new ship and its standard displacement in tons and metric tons, and the principal dimensions, namely, length at waterline, extreme beam at or below waterline, mean draft at standard displacement, at time of completion

(c) In case of loss or accidental destruction of capital ships or aircraft carriers, they may immediately be replaced by new construction subject to the tonnage limits prescribed in Articles IV and VII and in conformity with the other provisions of the present Treaty, the regular replacement program being deemed to be advanced to that extent.

(d) No retained capital ships or aircraft carriers shall be reconstructed except for the purpose of providing means of defense against air and submarine attack, and subject to the following rules: The Contracting Powers may, for that purpose, equip existing tonnage with bulge or blister or anti-air attack deck protection, providing the increase of displacement thus effected does not exceed 3,000 tons (3,048 metric tons) displacement for each ship. No alterations in side armor, in calibre, number or general type of mounting of main armament shall be permitted except:
  (1) in the case of France and Italy, which countries within the limits allowed for bulge may increase their armor protection and the calibre of the guns now carried on their existing capital ships so as not to exceed 16 inches (406 millimeters) and
  (2) the British Empire shall be permitted to complete, in the case of the *Renown*, the alterations to armor that have already been commenced but temporarily suspended.

Section II lists complete ship scrapping schedules for the five powers and is not included in this appendix.[1]

PART 4. Definitions

For the purposes of the present Treaty, the following expressions are to be understood in the sense defined in this Part.

Capital Ship

A capital ship, in the case of ships hereafter built, is defined as a vessel of war, not an aircraft carrier, whose displacement exceeds 10,000 tons (10,160 metric tons) standard displacement, or which carries a gun with a calibre exceeding 8 inches (203 millimetres).

Aircraft Carrier

An aircraft carrier is defined as a vessel of war with a displacement in excess of 10,000 tons (10,160 metric tons) standard displacement designed for the specific and exclusive purpose of carrying aircraft. It must be so constructed that aircraft can be launched therefrom and landed thereon, and not designed and constructed for carrying a more powerful armament than that allowed to it under Article IX or Article X as the case may be.

Standard Displacement

The standard displacement of a ship is the displacement of the ship complete, fully manned, engined, and equipped ready for sea, including all armament and ammunition, equipment, outfit, provisions and fresh water for crew, miscellaneous stores and implements of every description that are intended to be carried in war, but without fuel or reserve feed water on board.

The word "ton" in the present Treaty, except in the expression "metric tons," shall be understood to mean the ton of 2,240 pounds (1,016 kilos).

Vessels now completed shall retain their present ratings of displacement tonnage in accordance with their national system of measurement. However, a Power expressing displacement in metric tons shall be considered for the application of the present Treaty as owning only the equivalent displacement in tons of 2,240 pounds.

A vessel completed hereafter shall be rated at its displacement tonnage when in the standard condition defined herein.

CHAPTER III. MISCELLANEOUS PROVISIONS

Article XXI

If during the term of the present Treaty the requirements of the national security of any Contracting Power in respect of naval defence are, in the opinion of that Power, materially affected by any change of circumstances, the Contracting Powers will, at the request of such Power, meet in conference with a view to the reconsideration of the provisions of the Treaty and its amendment by mutual agreement.

In view of possible technical and scientific developments, the United States, after consultation with the other Contracting Powers, shall arrange for a conference of all the Contracting Powers which shall convene as soon as possible after the expiration of eight years from the coming into force of the present Treaty to consider what changes, if any, in the Treaty may be necessary to meet such developments.

Article XXII

Whenever any Contracting Power shall become engaged in a war which in its opinion affects the naval defence of its national security, such Power may after notice to the other Contracting Powers suspend for the period of hostilities its obligations under the present Treaty other than those under Articles XIII and XVII, provided that such Power shall notify the other Contracting Powers that the emergency is of such a character as to require such suspension.

The remaining Contracting Powers shall in such case consult together with a view to agreement as to what temporary modifications if any should be made in the Treaty as between themselves. Should such consultation not produce agreement, duly made in accordance with the constitutional methods of the respective Powers, any one of said Contracting Powers may, by giving notice to the other Contracting Powers, suspend for the period of hostilities its obligations under the present Treaty, other than those under Articles XIII and XVII.

On the cessation of hostilities the Contracting Powers will meet in conference to consider what modifications, if any, should be made in the provisions of the present Treaty.

Article XXIII

The present Treaty shall remain in force until December 31st, 1936, and in case none of the Contracting Powers shall have given notice two years before that date of its intention to terminate the treaty, it shall continue in force until the expiration of two years from the date on which notice of termination shall be given by one of the Contracting Powers, whereupon the Treaty shall terminate as regards all the Contracting Powers. Such notice shall be communicated in writing to the Government of the United States, which shall immediately transmit a certified copy of the notification to the other Powers and inform them of the date on which it was received. The notice shall be deemed to have been given and shall take effect on that date. In the event of notice of termination being given by the Government of the United States, such notice shall be given to the diplomatic representatives at Washington of the other Contracting Powers, and the notice shall be deemed to have been given and shall take effect on the date of the communication made to the said diplomatic representatives.

Within one year of the date on which a notice of termination by any Power has taken effect, all the Contracting Powers shall meet in conference.

Article XXIV

The present Treaty shall be ratified by the Contracting Powers in accordance with their respective constitutional methods and shall take effect on the date of the deposit of all the ratifications, which shall take place at Washington as soon as possible. The Government of the United States will transmit to the other Contracting Powers a certified copy of the *procès-verbal* of the deposit of ratifications.

The present Treaty, of which the French and English texts are both authentic, shall remain deposited in the archives of the Government of the United States, and duly certified copies thereof shall be transmitted by that Government to the other Contracting Powers.

IN FAITH WHEREOF the above-named Plenipotentiaries have signed the present Treaty.

DONE at the City of Washington the sixth day of February, One Thousand Nine Hundred and Twenty-Two.

    [SEAL] CHARLES EVANS HUGHES
    [SEAL] HENRY CABOT LODGE

[SEAL] OSCAR W UNDERWOOD
[SEAL] ELIHU ROOT
[SEAL] ARTHUR JAMES BALFOUR
[SEAL] LEE OF FAREHAM
[SEAL] A. C. GEDDES
R. L. BORDEN [SEAL]
   G. F. PEARCE [SEAL]
   JOHN W SALMOND [SEAL]
   ARTHUR JAMES BALFOUR [SEAL]
   V S SRINIVASA SASTRI [SEAL]
   A SARRAUT [SEAL]
   JUSSERAND [SEAL]
   CARLO SCHANZER [SEAL]
[SEAL] V. ROLANDI RICCI
[SEAL] LUIGI ALBERTINI
[SEAL] T. KATO
[SEAL] K. SHIDEHARA
[SEAL] M. HANIHARA

Appendix 2
U.S. Naval Policy, 1922*

Naval Policy is the system of principles, and the general terms of their applications, governing the development, organization, maintenance, training and operation of a Navy. It is based on and is designed to support national policies and American interests. It comprehends the questions of number, size, type, and distribution of naval vessels and stations, the character and the number of the personnel and the character of peace and war operations.

Fundamental naval policy of the United States
The Navy of the United States should be maintained in sufficient strength to support its policies and its commerce, and to guard its continental and overseas possessions.

U.S. Naval Policy Based on Treaty for Limitation of Naval Armament

Preface: The treat for the limitation of the naval armament, when promulgated, will be the supreme law of the powers party to the treat governing their naval armament as to battleships and aircraft carriers. The spirit of the treaty indicates two elements of international import: A general desire to avoid competition in naval armament. A partial condition of a ratio in naval strengths as a means of avoiding competition. Were any power now to undertake a program of expansion in unrestricted classes of naval vessels or in vessels not consistent with the treat ratios of capital ships, a new competition in naval strengths would thereby be initiated. Until such time as other powers by inequitable conduct in international rela-

---

*Note: NARA RG80 Proceedings and Hearings of the General Board, 17 January 1925, "Naval Policy." The December 1922 policy was attached to this 1925 hearing transcript. Its format, as much as possible, has been retained from the original.

tions as to U.S. interests or by their departure from the idea of a suspended competition in naval armaments, indicate other procedure, the Navy of the United States may be governed in naval strengths by the spirit of the capital shop rations, otherwise it will be necessary appropriately to adjust our naval policy.

General Naval Policy

To create, maintain, and operate a Navy second to none; and in conformity with the ratios for capital ships established by the treaty for the limitation of naval armament.

To make war efficiency the object of all training and to maintain that efficiency during the entire period of peace.

To develop and to organize the Navy for operations in any part of either ocean.

To make strength of the Navy for battle of primary importance.
To make strength of the Navy for exercising ocean-wide economic pressure next in importance.

To encourage, and endeavor to lead, in the development of the art and material of naval warfare.

Initiate friendly and sympathetic relations with the whole world by foreign cruises.

Support in every possible way American interests especially the expansion and development of American foreign commerce.

Maintain a Marine Corps of such strength that it will be able adequately to support the Navy by furnishing detachments to vessels of the fleet in full commission, guards for shore stations, garrisons for outlying positions; and by the maintenance in readiness of an expeditionary force.

Cooperate fully and loyally with all departments of the government.

Building and Maintenance Policy

To make the capital ship ratios the basis of building effort in all classes of fighting ships.

To make superiority of armament in their class an end in view in the design of all fighting ships.

To provide for great radius of action in all classes of fighting ships.

Capital Ships

To prepare and maintain detailed plans for new capital ship construction.

To replace existing capital ships in accordance with treaty provisions.

To keep all retained capital ship modernized as far as treaty terms permit, and good practice justifies.

To complete two ships of the West Virginia Class now building.

Aircraft Carriers

To convert now two battle cruisers to aircraft carriers, then: to build additional aircraft carriers at such a rate that the United States shall not fall behind treaty rations.

To prepare detailed type plans for immediate conversion of suitable merchant vessels to aircraft carriers.

To design aircraft carriers to carry as many combined torpedo, bombing and scout plans as possible.

Cruisers

To complete 10 light cruisers of the Omaha class now building.

To replace all old cruisers by building 16 modern cruisers of 10,000 tons displacement carrying 8 inch guns, and in addition to lay down and build cruiser tonnage at a rate that will maintain in cruiser strength the capital ship ratios.

Small Cruisers and Gunboats

To build no small cruisers.

To complete the gunboat now building.

To build in America especially fine, smart-appearing gunboats for the Chinese rivers and to build gunboats for no other purpose.

Destroyers

To complete those destroyers now building.

To maintain effective destroyer tonnage in conformity with capital ship ratios.

To lay down destroyer leaders first when it becomes necessary to undertake new destroyer construction.

To scrap no destroyer unless its material condition or its military characteristics make it undesirable for retention.

To make no further permanent structural changes in existing destroyers with a view to their assignment OT mine laying, scouting or other special operations.

Submarines

To complete submarines now building.

To maintain effective submarine tonnage in conformity with capital ship ratios.

To develop and build 12 scout submarines and 12 mine-laying submarines, and thereafter to limit new construction to that necessary to maintain effective submarine tonnage in conformity with capital ship rations.

To scrap no submarine unless its material condition or its military characteristics make it undesirable for retention.

Eagle Boats [Corvettes]

To build no more.

To retain all those that are in good material condition.

Sub-Chasers

To maintain only those necessary for training and experiment with listening devices.

Minelayers (Large)
To maintain 2 for training and development work.

Minelayers (Small)

To build no more.

To retain all those that are in good material condition.

Auxiliaries

To complete the auxiliaries now building.

To maintain the minimum number consistent with the training and the mobility of the fighting ships they serve.

To use commercial facilities for general transportation, where available, supplementing these by naval auxiliaries as may be necessary.

To prepare complete type plans for the quick conversion of suitable merchant vessels to serve as naval auxiliaries.

> #### Colliers, Oilers, Supply Vessels, Transports:
> To maintain those necessary under average conditions, chartering
> To meet special conditions.
>
> #### Tenders:
> To maintain one for each 15 destroyers in active service.
> To maintain one for each 38 destroyers in reserve.
> To maintain one for each 9 submarines not attached to bases and not less than 4 tenders,
> To maintain two for aircraft.

Fleet repair ships, target repair vessels, hospital ships, ammunition ships: To maintain 2 of each.

Cargo Vessels:
To maintain enough to supplement commercial carriers.

Tugs:
To maintain those for which there is active employment plus those necessary for rescue and salvage service.

Dispatch Boats:
10

Minesweepers:
To maintain sufficient for training and development, a minimum of 12.
To dispose of non-military vessels for which there is no prospective use and which can readily be obtained from commercial sources.

Air Policy

To complete rigid airships now under construction and to determine from their performance in serve the desirability of further construction.

To build non-rigid airships for development purposes only until their usefulness shall have been proved.

To direct the principal air effort on that part of the air service that is to operate from ships to the fleet.

To direct development of heavier than air craft principally towards spotting planes and towards torpedo, bomb and scout planes that can operate from ships. To combine the three latter functions in one plane.

To develop a combat plane for use afloat.

To acquire combat planes only as necessary for training and for current requirements.

To develop a scouting seaplane of long range for operations from a short base or from a tender.

To operate a spotting plane from each capital ship and from each modern cruiser.

To determine by trial the practicability and desirability of operating planes from destroyers and submarines.

To give every possible encouragement to aviation in civil life with a view to advancing the art and to providing aviators available for war.

Operating Policy[*]

To make and follow a general scheme of progressive education and training for the Navy.

Emphasize (?) in training that the principal elements of naval efficiency are morale, gunnery, and engineering in their tactical and strategical effects.

Deploy (?) the active fleet at least once a year for a period of not less than three months.

Keep capital ships fully manned and in active training.

Keep aircraft carriers full manned and in active training.

Keep modern cruisers fully manned and actively cruising.

Keep destroyers fully manned and in active training.

Keep submarines full manned and in active training.

Keep other modern fighting vessels partially manned and in commission.

---

[*]*Note*: The copy of this sheet in the microfilm obscured some of the language in this section and I have annotated those words that are unclear with a question mark (?). Organizational, Personnel, Allocation, Information, Base and Shore, Communications, Inspections, and Publicity Policy sections are not included in this transcript of the original.

Keep capital ships concentrated as far as possible. This not to preclude separate (?) divisional training periods.

Keep airplane carriers with the fleet:
   Their major mission with the battleships.
   Their secondary mission with scouts and other auxiliary arms of the fleet.

To put vessels assigned to reserve in condition for active service immediately and so to maintain them.

To mobilize vessels in reserve in rotation, at intervals, supplement their reduced crews with officers and men of the naval reserve.

To make foreign cruises as international conditions warrant to operate a naval train sufficient for the upkeep and mobility of the fighting ships an expeditionary forces it serves, and designed to be the nucleus of a transport and trains service in case of war.

To make every effort both ashore and afloat, at home and abroad, to assist the development of the American interests and especially the American Merchant Marine.

To make every effort for economy in expenditures while holding efficiency paramount.

To assign suitable, partially manned vessels for the training of naval reserves.

Appendix 3

Bureau Recommendations for Battleship Modernization*

Bureau of Engineering
    a. Boilers and propelling machinery, involving possible change in type of machinery (meaning replacement of reciprocating engines with turbo-electric drive).
    b. Conversion of coal burners to oil burners
    c. Electric light plant and communication

Bureau of Ordnance
    d. Battery (gun turret changes, unspecified as yet)
    e. Torpedo armament
    f. Fire control

Bureau of Construction and Repair
    g. Torpedo protection. This is referred to more fully in paragraph 6 below.
    h. Deck protection.
    i. Principal (mechanical) auxiliaries, such as steering gear, anchor gear, and so on.

Bureau of Aeronautics
    j. Airplane equipment

6. One of the most important of these items is that of torpedo and deck protection. The Treaty allows an increase in displacement of 3,000 tons for the purpose of

---

*Note: From GB Serial No. 1121, 26 April 1922, attached joint letter of 04 April 1922 from BuEng, BuAer, BuC&R, and BuOrd.

equipping existing tonnage with bulge or blister or anti-air attack deck protection. Now in the case of at least some of our existing ships, it appears that greatly improved torpedo protection can be given by adding bulkheads within the present coal bunkers or other <u>internal</u> spaces; that is, the additional protection sought can be fitted <u>inside</u> the ship. The question at issue here then is, does the Treaty permit additional torpedo protection to be provided for <u>inside</u> the ships, or must this be added as a bulge or blister <u>outside</u> the existing ship [underline in original].

Note: The General Board answered this question conclusively in Serial 1121 saying that "unlimited alterations" were authorized inside the ship as long as tonnage was not increased.[1]

Appendix 4

Comparison Chart for Cruiser Designs[1]

26 December 1930.

Tabulation of Aviation Facilities Afforded by Designs of 10,000 Ton Cruisers with Landing-on Deck Submitted to the General Board by the Bureau of Construction and Repair:

II	III	IV	V	VI	VII	VIII
Type	Capacity of Flight Deck with Planes Spotted for Take-off into 30 m.p.h. Wind	Capacity of Hangar for Planes Stowed with Wings Spread	Capacity of Flight Deck for Planes Stowed Leaving 250' Arresting Gear Clear	Maximum Operating Capacity Equals Column IV Plus Column V	Recommended Operating Capacity Total Taken as the Multiple of Six Which is Equal to or Less than the Total in Column IV	Time of Launching the Number of Planes Indicated in Column VII. Time Interval for Fly-off is 10 Seconds, or Catapulting 5 Minutes, both Intervals now in use in Fleet.
Side island, one elevator forward	VF type – 14 VO type – 13 Total 27	VF type – 10 VO type – 9 Total 19	VF type – 6 VO type – 5 Total 11	VF type – 16 VO type – 14 Total 30	VF type – 9 VO type – 9 Total 18	3 minutes
Side island, one elevator forward	VF type – 14 VO type – 14 Total 28	VF type – 11 VO type – 11 Total 22	VF type – 6 VO type – 5 Total 11	VF type – 17 VO type – 16 Total 33	VF type – 9 VO type – 9 Total 18	3 minutes
Side island, one elevator forward	VF type – 15 VO type – 15 Total 30	VF type – 11 VO type – 12 Total 24	VF type – 7 VO type – 7 Total 14	VF type – 19 VO type – 19 Total 38	VF type – 12 VO type – 12 Total 24	4 minutes
Flush deck aft, one elevator forward	VF type – 13 VO type – 13 Total 26	VF type – 11 VO type – 11 Total 22	VF type – 4 VO type – 3 Total 7	VF type – 15 VO type – 14 Total 29	VF type – 9 VO type – 9 Total 18	3 minutes
Flush deck forward, one elevator aft	VF type – 13 VO type – 13 Total 26	VF type – 8 VO type – 7 Total 15	VF type – 4 VO type – 3 Total 7	VF type – 12 VO type – 10 Total 22	VF type – 6 VO type – 6 Total 12	2 minutes
Center island, one elevator forward, 2 catapults in hangar	VF type – 10 VO type – 10 Total 20* *Note: this would require backing the ship to obtain a 30 m.p.h. wind. Take-off toward stacks impossible for squadron operation.	VF type 5 + 1 (cat.) = 6 VO type 4 + 1 (cat.) = 5 Total 11	VF type – 0 VO type – 0 Total 0	NOTE: Planes in squadron units could not be landed with safety on this arrangement, Design G.	VF type – 6 VO type – 6* Total 12* *Note: Based on assumption that one additional VO type can be carried.	2 minutes if flown off over stern. *25 minutes by catapults. *Note: Further development of catapult equipment and procedure might reduce this time by 50%.

Appendix 5

Excerpt from the Mobile Base Project*

(b) Docking Facilities

10. The assumptions in the preceding paragraph 7(a) will require the following docking facilities:

Class A—2 mobile floating drydocks to accommodate vessels above 20,000 tons displacement.

Class B—2 mobile floating dry docks to accommodate vessels from 12,000 to 20,000 tons displacement.

Class C—4 mobile floating dry docks to accommodate vessels from 3000 to 12,000 tons displacement.

Class D—11 mobile floating dry docks to accommodate vessels below 3000 tons displacement.

Note: (a) The use of caissons will reduce the above numbers of docks to the extent of their practicability for emergency repairs.

(b) The above docking facilities are designed to accommodate the following vessels:

(1) All ships of Appendix B –Basic Readiness Plan –less Special Service Squadron, and including Converted Combatant Types as follows:

---

*Note: RG80, Chief of Naval Operations Correspondence, 198-1, 1919–1927, Appendix F to WPL-9, December 1923, 10-11. Format is from the original.

9 XOCV
13 XOCA
4 XOCL
9 XOCM

(2) Naval Transportation Service Vessels in current maintenance service, which are assumed to be 90 ships per month.

(c) These vessels are assumed to receive docking facilities as follows:

(1) All combatant types plus all AD, AS, AG, AR, AT, AP, AND AV, both routines and emergency docking.

(2) All AE, AH, AK, AF, AO, and Naval Transportation service vessels in current maintenance Service, emergency docking only.

(3) <u>Routine Docking</u>. All combatant ships 6 months routine docking period; all others, which receive routine docking, 12 months routine docking period.

BB, CV, XAP, and XOCV - 1 ship per dock, 7 days in dock.
DD, DM, SS, AT, and PG - 2 ships per dock, 3 days in dock.
All others - 1 ship per dock, 4 days in dock.

(4) <u>Emergency Docking.</u>

All combatant ships, less DD, DM, SS and AM	- 2% casualties
DD, DM, SS, and AM	- 3% casualties
N.T.S. on current maintenance	-2% casualties
All others	-1% casualties

(e) Each class of dock, working continuously, will accommodate ships as follows, on the above assumptions:

Class A—36 ships
Class B—107 ships
Class C—104 ships plus 90 ships per month of N.T.S.
Class D—<u>496 ships</u>
Total—743 ships plus 90 ships per month of N.T.S.

Notes

Preface

1. Captain Dudley W. Knox, *The Eclipse of American Sea Power* (New York: American Army and Navy Journal, Inc., 1922), 102–103.

2. William F.Trimble, "Admiral Hilary P. Jones and the 1927 Geneva Naval Conference," *Military Affairs*, Vol. 43 (Feb. 1979), 1–4. Trimble attributes the this characterization of Admiral Jones to historian Benjamin H. Williams and it has ever since been parroted by a broad range of naval historians.

Chapter 1

1. My use of the terminology "treaty system" conforms to that of Thomas H. Buckley in "The Washington Naval Limitation System: 1921–1939," in Richard Dean Burns ed., *Encyclopedia of Arms Control and Disarmament, Volume II* (New York: Charles Scribner's Sons, 1993).

2. See Waldo H. Heinrichs Jr. "The Role of the United States Navy," in Dorothy Borg and Sumpei Okamoto, eds. *Pearl Harbor as History: Japanese-American Relations, 1931–1941* (New York: Columbia UP, 1973); Peter Karsten, *The Naval Aristocracy: The Golden Age of Annapolis and the Emergence of Modern American Navalism* (New York: The Free Press, 1972); Robert L. O'Connell, *Sacred Vessels: The Cult of the Battleship and the Rise of the U.S. Navy* (Boulder, CO: Westview Press, 1991); and William M. McBride, *Technological Change and the United States Navy, 1865–1945* (Baltimore MD: The Johns Hopkins UP, 2000).

3. George T. Davis, *A Navy Second to None: The Development of American Naval Policy* (New York: Harcourt Brace & Co., 1940), 275–276.

4. "Washington Naval Treaty" refers to the Naval Arms Limitation Treaty signed at Washington in 1922. This treaty is often simply referred to as the Washington Naval Treaty. Many treaties were signed at Washington in 1922 and this usage identifies the specifically naval treaty to avoid any confusion with the other treaties and agreements.

5. See Norman Friedman, Thomas C. Hone, Mark Mandeles, *America and British Aircraft Carrier Development, 1919–1941* (Annapolis: Naval Institute Press, 1999);

Charles Melhorn, *Two-Block Fox: The Rise of the Aircraft Carrier, 1911–1931* (Annapolis: Naval Institute Press, 1974); and Clark Reynolds, *The Fast Carriers: The Forging of an Air Navy* (New York: McGraw-Hill, 1968).

    6. For the latest discussion of sea-basing and the U.S. Navy see Geoffrey Till, *Naval Transformation, Ground Forces, and the Expeditionary Impluse: The Sea-Basing Debate* (Carlisle Barracks, PA: Army War College, Government Printing Office, 2006).

    7. James S. Corum, *The Roots of Blitzkrieg: Hans von Seeckt and German Military Reform* (Lawrence KS: University Press of Kansas, 1992), passim. See also Williamson Murray, "Armored warfare: The British, French, and German experiences," in W. Murray and A. R. Millet, eds., *Military Innovation in the Interwar Period,* Cambridge: Cambridge University Press, 1996, 6–49.

    8. For an articulate presentation of this view see Emily Goldman, *Sunken Treaties: Naval Arms Control Between the Wars* (University Park, PA: The Pennsylvania State University Press, 1994), 31.

    9. Williamson Murray, "Innovation past and future," from W. Murray and A. R. Millett, eds. *Military Innovation in the Interwar Period* (Cambridge: Cambridge University Press, 1996), 311.

    10. Friedman, Hone, and Mandeles, *American and British Aircraft Carrier Development*, 5–6. These authors cite Nobel Laureate Douglass C. North, *Institutions, Institutional Change, and Economic Performance* (New York: Cambridge University Press, 1993.) More recently, Mandeles acknowledges the role of Jean de Bloch's pioneering analyses in *The Future of War: Organizations as Weapons* (Washington, D.C., 2005) as contributing to his inspiration for the "levels of analysis" approach.

    11. Alan Beyerchen, "From radio to radar: Interwar military adaptation to technological change in Germany, the United Kingdom, and the United States," from W. Murray and A. R. Millet, eds., *Military Innovation in the Interwar Period* (Cambridge: Cambridge University Press, 1996), 268 and passim.

    12. William R. Braisted, "The Evolution of the United States Navy's Strategic Assessments in the Pacific, 1919–1931." in Erik Goldstein and John Maurer, eds. *The Washington Conference, 1921–1922: Naval Rivalry, East Asian Stability and the Road to Pearl Harbor,* (Illford, U.K.: Frank Cass, 1994), 102. The first attempts had occurred prior to World War I and obviously had failed, see also Michael Howard, *War and the Liberal Conscience* (New Brunswick, NJ: Rutgers University Press, 1994).

    13. For analysis of the interwar treaties as a "system" see Buckley "The Washington Naval Limitation System: 1921–1939," 639.

    14. Thomas S. Kuhn, *The Structure of Scientific Revolutions* (Chicago: University of Chicago Press, 1962), passim.

    15. Sir Julian Corbett, *Some Principals of Maritime Strategy.* Reprint (Annapolis: Naval Institute Press, 1988), p.336–337. The term "balanced fleet" is not specifically used, but rather the concept is that of a fleet balanced to achieve the objectives set before it, e.g. protecting communications, attacking enemy communications, and attacking the enemy fleet.

16. The period effectively ended after the conclusion of the Second London Naval Conference in 1936, although its formal demise came later. Some writers identify the end of the treaty system as 1938 when the United States and Great Britain used the so-called "escalator" clauses to build over treaty limits. See Chapters 2 and 4. The escalator clauses essentially let a signatory build in excess of limits if one of the original signatories built in excess of treaty limits. See also Appendix 1.

Chapter 2

Note: The epigraphs were taken from Waldo H. Heinrichs, Jr. "The Role of the United States Navy," in Dorothy Borg and Sumpei Okamoto, eds. *Pearl Harbor as History: Japanese-American Relations, 1931–1941* (New York: Columbia University Press, 1973), pp 200–201; Norman Friedman, Thomas C. Hone, & Mark Mandeles, *America and British Aircraft Carrier Development, 1919–1941* (Annapolis: Naval Institute Press, 1999), 191.

1. Philip L.Semsch, "Elihu Root and the General Staff," *Military Affairs* 27, No. 1 (Spring 1963), 16–27.

2. The Naval Historical Center at http://www.history.navy.mil/faqs/faq65-1.htm, accessed 13 October 2006.

3. Gerald E. Wheeler, *Admiral William Veazie Pratt, U.S. Navy: A Sailor's Life* (Washington, D.C.: Naval Historical Division, Department of the Navy, 1974), 67–71. Wheeler argues that the Navy used the course of instruction at the Naval War College to weed out the "fools" and identify the "brilliant."

4. See Heinrichs, Philip Crowl, and Robert Love. It was only in the 1970s after the wholesale declassification of U.S. archival material up through World War II that historians had complete access to the complete records of the General Board. It should be no surprise, then, that the role of the General Board was not well understood up through the early 1970s—almost on par with how it was viewed in its own day.

5. George Baer, Edward Miller, William Braisted, Mark Melhorn, Thomas C. Hone, et al. all use the General Board's archival records to support and advance their arguments. Of the works cited, Braisted and Hone et al. have done the most to advance our understanding of the General Board in recent years.

6. Harlow A. Hyde, *Scraps of Paper: The Disarmament Treaties Between the Wars* (Lincoln, NE: Media, 1988), 193. Hoover's Armistice Day 1929 speech proclaimed, "We will reduce our naval strength in proportion to any other. . . . It cannot be too low for us."

7. Hoover Library (HL). Box 156, Memo from General Board (GB) to Assistant Secretary of the Navy dated 14 December, 1929. Hereafter HL, GB memo December 14, 1929.

8. Ronald Spector, *Professors at War: The Naval War College and the Development of the Naval Profession* (Newport, RI: Naval War College Press, 1977), 143. Spector makes a direct link between organizational reform in the Navy as a subset of the larger societal trends and developments, especially the Navy's quest for "efficiency" and the Navy's "business approach to war." See also Robert H. Wiebe, *The Search for Order, 1877–1920* (New York: Hill and Wang, 1967).

9. Ronald W. Knisley, *The General Board of the United States Navy—Its Influence on Naval Policy and National Policy,* Thesis (M.A.) (University of Delaware, 1967), 12. This board was dissolved after 1898. Knisley's work only covers the period through 1921 but runs out of archival data for the General Board circa 1917.

10. Robert W. Love, *History of the U.S. Navy,* Vols. I and II, 1775–1991 (Harrisburg, PA: Stackpole Books, 1992), 417. HL, GB memo December 14, 1929, 2.

11. HL GB memo, December 14, 1929. National Archives and Records Administration (NARA) Proceedings and Hearings of the General Board of the U.S. Navy, 1900–1950 (hereafter PHGB), Roll 1 archivist comments.

12. As mentioned, the creation and subsequent reorganizations of the General Board occurred before the famous reforms of Elihu Root under President T. Roosevelt. Root later served as a principal delegate and to the Washington Conference.

13. HL GB memo, December 14, 1929. NARA PHGB, roll 1, archivist's comments.

14. The Bureau of Ordnance was nicknamed the "gun club" due to its advocacy of all calibers of guns for ships and other ordnance such as torpedoes. It also made recommendations for armor for the various classes of ships and tended to be the home of the fiercest partisans of the battleship.

15. Unrestricted line officers could command ships at sea whereas restricted line or staff officers such as the CC officers could not. The CC officers were intimately involved in ship design and were transformed by World War II into what they are today, a base and port civil construction corps also known as the "Seabees." Ship design oversight migrated to Chief of Naval Operations and today resides within Chief of Naval Operations "NAVSEA" (Naval Sea Systems Command).

16. Knisley, 8–9. Knisley highlights the "jealously independent" character of the separate bureaus prior to the establishment of the GB. See also Love, 417.

17. John Hood, "Naval Administration and Organization," *United States Naval Institute Proceedings,* Vol. 27, No. 1 (March 1901), 1 and subsequent comments 30; PHGB, minutes 23 April 1901. Hood discusses a "General Staff" but nowhere mentions the General Board. In the subsequent discussion commentator Capt. F. E. Chadwick does mention the General Board "acting as a balance-wheel upon the undue zeal of self-centered projects." Chadwick was recommending in these comments a change in the composition of the General Board to a smaller group that met more " . . . frequently, if not daily . . . ". The General Board's minutes of April 23, 1901 simply records the change effected by General Order 43.

18. HL GB memo, December 14, 1929. This memo puts the term "aid" in quotes and does not spell it "aide" as it usually was (and still is) for the secretary's assistants.

19. Ibid. The term "ex-officio" is reproduced here as written throughout the memorandum.

20. NARA PHGB, Roll 1 archivist comments. HL, GB memo December 14, 1929, 2.

21. Part of the difficulty here is that what we today call the operational level was in that day and age considered strategic because the concept of an operational level of war was not yet well developed in U.S. military doctrine or usage. In the Army's Field Service Regulations of 1923 only the tactical and strategic levels are mentioned. By 1939 the Army's new Field Service Regulation was FM 100-5 entitled *Operations* still did not men-

tion the operational level of war; operations were either strategic or tactical. The Navy used Navy Regulations and naval warfare publications for its doctrine. These also do not talk about the operational level. It was only after the Vietnam War that U.S. service doctrine began to talk about the operational level of war.

22. HL, GB memo December 14,1929, 2.

23. Ibid., 4.

24. PHGB, July 3, 1931 minutes. Pratt's action in making this change has sometimes been regarded as an action taken to reduce the influence of the Board, for example see Scott T. Price, "A Study of the General Board of the U.S. Navy, 1929–1933," unpublished master's thesis (Omaha, NE, 1989), 134. However, the evidence is clear that Pratt wanted the Board's independent advice to be free from the influence of the Chief of Naval Operations, at that time a member of the Board. It also meant the Chief of Naval Operations could freely give his advice and comment without contradicting the opinion of a body he also served on. See Wheeler, 323–324 and PHGB, June 30, 1931, "Recommendations of Changes in U.S. Naval Regulations 1920," for the pertinent hearing on the subject by the General Board that included testimony by Pratt.

25. George W. Baer, *One Hundred Years of Sea Power: The U.S. Navy, 1890–1990* (Stanford, CA: Stanford University Press, 1971), 301. Baer states that the disestablishment of the Board by Secretary of the Navy Mathews in 1951 "was a loss to the service."

26. PHGB Roll 1 membership list (hereafter PHGB ML), 1939–1940. King reported to the General Board in August 1939 and detached in December 1940 to become Commander-in-Chief of the Atlantic Fleet, a combat assignment given the ongoing Battle of the Atlantic, then Commander-in-Chief of the U.S. Fleet (COMINCH), and finally chief naval operator in March 1942. Note how King changed CINCUS to COMINCH.

27. NARA RG80, 438-1, February 17, 1922 GB request to Professor G. Wilson of Harvard for opinion on Washington Treaty technical legal matter. PHGB, 1925, 888. Dr. W. F. Durand of Stanford University and a member of the National Advisory Board on Aeronautics testified before the Board with regard to a unified air service not long after General "Billy" Mitchell.

28. PHGB ML, 1919–1940.

29. NARA RG80, GB 449, January 27, 1931 Chief of Naval Operations memorandum, "Naval Air Operating Policy." Pratt uses the term "authoritative" in referring to the promulgated policy, which had been written by the General Board. The U.S. Army uses the word "authoritative" to refer to the role that military doctrine plays in conducting operations. See FM 3-0 *Operations* (Washington, D.C.: U.S. Government Printing Office, 2001), page 1–14.

30. GB 449 War Department, October 21, 1919; GB 420-2 Secretary of the Navy, 07 September 1928; Curtis D. Wilbur,"To Authorize Major Alterations to Certain Vessels," (H.R. 8353), 27 March 1924; GB 438 Secretary of the Navy, January 18,1927; Correspondence of the Chief of Naval Operations, July 11, 1936; GB 420-2, Chief of Naval Operations September 9,1941. These are all representative examples of correspondence that refer to the authoritative nature of the General Board's advice.

31. The General Board did produce minutes (which the archives term "proceedings") for every single meeting, but these were usually no more than one page and did not include

transcripts of discussions. For example, the minutes for March 3, 1922 listed the membership in attendance, read and approved the minutes of the previous meeting and simply stated that "The Committee discussed Naval Policy" and gave the names of the transcribers for the hearing held on that topic. It is often difficult from these minutes to tell if a hearing was held at all since they rarely discuss the content of the actual meeting. The minutes will often simply state that discussion of a particular issue began at a certain time and ended at a certain time.

32. HL GB memo, December 14, 1929, 4.

33. PHGB, Roll 1, Index of General Board Serials and archivist comments. The assignment of the serial numbers remained chronological, but later records of the General Board dropped the date of receipt and simply listed the date the finalized serial was submitted to the secretary.

34. NARA RG80 GB 438-1, 09 July 1929 letter from the Secretary of State to the Secretary of the Navy proposing a new "yardstick" method for arms limitation, this caused the Navy secretary to consult the General Board.

35. See NARA GB 420-2 Serial No. 1083. Serial 1083 was the result of inputs requested by the Board for a Navy Building Policy, an agenda item stimulated from the "bottom up." Extensive hearings on this initiative occurred in 1923; see NARA PHGB 1923 "Building Policy." Appendix II is the U.S. Naval Policy approved in December 1922.

36. A good example of this was Hoover's direct communications with the Board and Adm. Hilary Jones in attempting to craft a new formula for a "naval yardstick" prior to the London Naval Conference in 1930. NARA, GB 438-1, September 28, 1929 letter from the president to the General Board, in same folder as Serial 1449 as background for the Board's response.

37. NARA RG80 PHGB, March 3,1922, "Naval Air Policy," 26–57. Note, PHGB page numbers are microfilm page numbers, not the actual pages of the transcripts.

38. See NARA RG80 GB 449, October 21, 1919, War Department Memo. This memorandum from the War Department to the Secretary of the Navy acknowledged that Mitchell had erred in concluding the General Board agreed that air power had made "navies almost useless." For Mitchell's original testimony see PHGB, April 3, 1919, for the restrained tone and curious interest evinced by the General Board in the face of some of Mitchell's more extravagant claims. The Board's consideration of new ideas over the long term is better illustrated by the hearings and history of the flying deck cruiser (see Chapter 6).

39. PHGB, NARA archivist comments, 2–3. NARA RG80, GB 449, Serial 1140, November 18, 1922 and 1st and 2nd endorsements, passim.

40. General Board serials were typically written in a three part format: introduction, discussion, recommendations.

41. NARA RG80, Serial No. 1369 is one example of a finalized serial with marginalia referring to minutes, previously staffed copies (with different dates) and instructions to hold for possible revision as a differently numbered serial.

42. PHGB, December 4, 1930. The hearings on the flight deck cruiser are rich with references to design compromises (see chapter 6).

43. NARA RG80 GB 438-1 Serial No. 1444-A, September 11, 1929 memorandum. This memo was attached to the serial and reflects an extended session where the members

of the General Board clearly compromised in order to come up with a suitable cruiser policy for the 1930 London Naval Conference (see Chapter 4).

44. The General Board provided the Secretary recommendations for the budget two years in advance of a particular fiscal year. The same process is still in place today.

45. NARA RG80, GB 449, Serial 1140, November 18, 1922 and 1st and 2nd endorsements, passim and PHGB, Proceedings Report, November 28, 1922, 254. The "U.S. Naval Policy" that reflected this serial was published 01 December, 1922 (see Appendix 2).

46. In 2006 the U.S. Navy refers to "warfare communities" in the same way that the U.S. Army more formally refers to combat arms, combat support, and combat service support "branches."

47. Evidently the use of the term "balance wheel," which today might seem archaic to anyone but a mechanical engineer, was very common in Pratt's day and earlier. He did not have to explain its meaning to anyone at the rather well-attended hearing where he coined the metaphor. See also the use of this term in connection with the General Board in *The Proceedings of the United States Naval Institute*, Vol.. 27, No. 1 (March 1901), 30.

48. Baer, 135, 139–141; Ronald H. Spector, "The Military Effectiveness of the US Armed Forces, 1919–1939," in *Military Effectiveness, Volume II The Interwar Period*, Eds. Allan R. Millett and Williamson Murray (Boston: Unwin Hyman, 1988), 90; and Rosen, Stephen Peter. *Winning the Next War: Innovation and the Modern Military*. Ithaca: Cornell University Press, 1991, 222–223.

Chapter 3

Note: The epigraph is taken from NARA RG80 GB, "Some Effects of the Washington Conference on American Naval Strategy," by Capt. F. H. Schofield, lecture delivered at the Army War College, Washington Barracks, September 22, 1923.

1. Clark Reynolds, *Command of the Sea* (New York: William Morrow & Co., 1974), 5–7. Reynolds' first chapter reflects the view of the sailor as both enamored of technology while at the same time of a conservative mindset.

2. See Samuel Huntington, *The Soldier and the State* (Cambridge, MA: Harvard University Press, 1957), Chapter 4 passim. Huntington's classic remains the classic argument in support of this point: that modern militaries are by their very nature conservative.

3. Elting E. Morison, "The Navy and the Scientific Endeavor," in *Science and the Future Navy: A Symposium* (Washington, D.C.: National Academy of Sciences, 1977), 14.

4. Philip L. Semsch, "Elihu Root and the General Staff," *Military Affairs 27*, No. 1 (Spring 1963), 16–27. The Root Reforms were named after Secretary for War Elihu Root and led to the establishment of the Army War College, a planning staff and a chief of staff for the Army by the end of Theodore Roosevelt's presidency. Root was later a delegate to the Washington Naval Conference. See Ronald Spector, *Professors at War*, passim and George Baer, *One Hundred Years of Sea Power*, Chapter 3, passim.

5. H. P. Willmott, *The Battle of Leyte Gulf* (Bloomington, IN: Indiana University Press, 2005), 3. See also James Hammond, *The Treaty Navy* (Victoria, Canada: Trafford, 2001), 174. Both Hammond and Willmott agree that the U.S. Navy built during the interwar period

fought the Japanese Navy to a standstill, which effectively defeated its prewar strategy to get the United States to negotiate an early peace.

6. The fixed ratio, which granted the U.S. Navy superiority in capital ships and the right to maintain superiority in other categories, offset the "advantage" it was felt the Japanese gained through the fortification clause.

7. See Phillip A. Crowl, "Alfred Thayer Mahan: The Naval Historian" in P. Paret, ed., *Makers of Modern Strategy* (Princeton, NJ: Princeton University Press, 1986), 474–476. Crowl is chief among those who claim that Mahan, even in his own time, no longer influenced the way the Navy operated and thought. Historians Stephen Roskill, Steven E. Pelz, and Sadao Asada take the opposite view, attributing the strategic ideas of all three major navies to Mahan. W. Stephen Roskill, *Naval Policy Between the Wars*, Vol.. 2, *The Period of Reluctant Rearmament, 1930–1939* (Annapolis, MD: Naval Institute Press, 1976), 174; Stephen E. Pelz, *Race to Pearl Harbor* (Cambridge, MA: Harvard University Press, 1974), 25–26, 88; Sadao Asada, "From Washington to London," in E. Goldstein and J. Maurer, eds., *The Washington Conference, 1921–1922* (United Kingdom: Frank Cass, 1994), 150.

8. Senate Document No. 77, 11–17. These pages encompass the text of Secretary Hughes' opening speech. See also Thomas Buckley, *Encyclopedia of Arms Control and Disarmament, Vol.* II (New York: Charles Scribner's Sons, 1993), 642–643.

9. Sadao Asada, "Japanese Admirals and the Politics of Naval Limitation: Kato Tomosaburo versus Kato Kanji," in Gerald Jordan, ed., *Naval Warfare in the Twentieth Century, 1900–1945: Essays in Honor of Arthur Marder* (New York: Crane Russak, 1977), 141, 150–156. Asada claims that Baron Kato "decided at once" to accept the American proposal and that the problem of the ratio had to do with internecine strife within the delegation and the Naval Ministry back in Japan.

10. A. T. Mahan, *Naval Strategy* (Newport, RI: Department of the Navy, GPO, 1991). Originally published 1909, reprint 1991 as U.S. Marine Corps pub *FMFRP 12–32*, 160–163.

11. George W. Baer, *One Hundred Years of Sea Power*, 95.

12. Sadao Asada, "From Washington to London," 149–150.

13. Ibid., 153.

14. Senate Document No. 126, *Conference on the Limitation of Armament* (Washington, D.C.: GPO 1922), 252. This document includes the proceedings of the plenary sessions of the conference. The final form of Article XIX is not much different than that captured by the proceedings. See also NARA RG80 GB 438-1, Address by Rear Admiral H.S. Knapp, April 27,1922, 7. See also Appendix 1.

15. Edward Miller, *War Plan Orange* (Annapolis: Naval Institute Press, 1991), Chapters 10–13 passim. Miller shows in these chapters how slow Navy planners were to respond to the strategic realities imposed by Article XIX.

16. Asada, 155–156. Baron Kato's principal opponent was Vice-Admiral Kato Kanji (no relation), a "hot blooded" member of the naval general staff assigned to the delegation as a technical advisor.

17. NARA RG80 GB 420-2 Serial No. 1083, July 15, 1921. The General Board had recommended building up Guam and improving the basing facilities in the Philippines in

1921, however this recommendation had been shelved due to both cost and concerns by the Harding administration that it would provoke the Japanese.

18. Asada, 155. Kato Tomosaburo would go on to become Prime Minister of Japan although he would die not long after the great Tokyo earthquake of 1923. His policy of friendship with the United States would not last due to American immigration laws and his own untimely death in 1923.

19. Roger Dingman, *Power in the Pacific*, 210–212.

20. Buckley, 643.

21. NARA RG80 GB 438-1, January 6, 1922, "Minutes of the Tenth Meeting of the Advisory Committee." These minutes were prepared for the American delegation at the Washington Naval Conference.

22. Buckley, "The Washington Naval Limitation System," 645.

23. NARA RG80 GB 420-2, Chief of Naval Operations "Memorandum in Re Tentative Draft of U.S. Naval Policy," March 4, 1922, 1. The Open Door policy refers to the policy of equal economic opportunity and markets for all the major powers in their trade relations with China.

24. Dudley W. Knox, "The Role of Doctrine in Naval Warfare," *United States Naval Institute Proceedings,* vol. 41, No. 2 (March–April 1915), 349. Beginning in 1918 Knox served in the historical section of the OpNav staff as its head until 1946.

25. Mahan, *The Influence of Sea Power Upon History* (New York: Hill and Wang, 1957), 75.

26. Mahan, *Naval Strategy,* 110.

27. Ibid., 122.

28. GB 420-2, Serial 1251, November 3, 1924.

29. Baron Antoine Jomini was author of many famous treatises on the art of war and has been accused, by Clausewitz among others, of reducing war to simple formulas and principals. See John Shy, "Jomini," in *Makers of Modern Strategy*, ed. Peter Paret (Princeton, NJ: Princeton University Press, 1986), 143.

30. A. T. Mahan, "The Battle of the Sea of Japan," USNI *Proceedings*, Vol. 23 No. 2 (June 1906), 450–452.

31. B. A. Fiske, "The Relative Importance of the Philippines and Guam," Special Supplement to the *United States Naval Institute Proceedings*, No. 225 (November 1921), 1676-1.

32. Michael Vlahos, *The Blue Sword* (Newport, RI: Naval War College Press, 1980), i–v. NARA RG38 Naval War College Problems Series IIA, Strategic Problem IV, February 9, 1926 debrief transcript. Note that the Navy entitled this particular exercise, a Philippine scenario in an War Plan Orange, as a "Strategic Problem." See also Spector, *Professors at War*, passim.

33. William S. Sims, "The United States War College," *United States Naval Institute Proceedings*, vol. 45, no.9 (September 1919), 1490. The College had temporarily suspended class during World War I. The actual address was delivered to the new class in June 1919.

34. RG80, Correspondence of the office of the chief of naval operations, PD 198-1, October 21,1919, "Statement of Admiral McKean." This memorandum highlights McKean's initiative to link war plans assessment with Naval War College gaming. McKean

was the Assistant to the chief of naval operations. Jeffrey Barlow, email, Wednesday 9, 2006, to author, source: officer biography Adm. Josiah McKean, Naval Historical Center.

35. NARA RG38, Records of the War Plans Division, office of the chief of naval operations, 1912–1946. WPL-9, vol. II, Part Three, 14–24. These pages of Plan Orange especially frame the missions and subordinate missions. For NWC wargaming of this plan see RG38, War Plans Division, OpNav, NWC Problems Series II-A, Joint Army and Navy Problem 1922–1923, NWC-AN3/GSS–EF7 (published by GSS, Ft. Leavenworth, KS), 1–8. The Mission was to recapture Philippines and the Estimate of the Situation followed.

36. PHGB, March 3, 1922, "Naval Air Policy." This hearing included a lengthy discussion over the issue of numbers of airplane tenders, an innovative new vessel, for the Pacific Fleet among Captain Schofield, Admiral Pratt, and Captain Mustin. It highlights the general awareness, before ratification of the treaty, of the impact of the fortification clause on naval operations and logistics in the Pacific.

37. Knox retained this position through World War II. His function was almost an exact analog of the historical section of the German General Staff. NHC still works for the Chief of Naval Operations today and its Director is a senior active-duty Navy officer.

38. Knox, 102–103, 135.

39. Ibid., 58–59, 54–56. PHGB, December 3, 1923, "Characteristics of Floating Drydocks," also emphasizes the rapidity of the Orange Plan's rewrite to include a mobile base project to ameliorate the basing problem for the fleet. The mobile base project had received chief of naval operation's blessing by December 1922.

40. Miller, *War Plan Orange*, Chapter 10, passim. Miller argues that the fortification clause initially eliminated what he called the "through ticket" thrust to bases in the Philippines if war broke out with Japan as a viable Orange course of action in war planning.

41. NARA RG80 GB 438-1, Knapp, 27 April 1922, 6–7.

42. GB 438-1, Transcript of address of Charles Evans Hughes December 28, 1922, 26–27.

43. W.V. Pratt, "Some Considerations Affecting Naval Polity," *United States Naval Institute Proceedings* (November 1922), 1845–1862. See also Wheeler, 185–186. Wheeler lists the article as "policy" when in fact the word that Pratt uses, and emphasizes, is "polity"—Pratt is speaking directly to the officer corps as a political component of the larger American polity. Pratt also wrote articles supporting the Treaty in *The North American Review* (1922) and *Current History* (1923).

44. NARA RG80 PHGB, March 3, 1922, "Naval Air Policy," 26–57. Note, PHGB page numbers are microfilm page numbers, not the actual pages of the transcripts.

45. PHGB, Proceedings, March 3, 1922 and hearing transcripts for that date. PHGB Roll 1 Members List. Note: the minutes for the hearings and the actual transcripts are located in two different sections of the RG80 microfilm record (1493) for the year 1922.

46. PHGB, March 3, 1922, 26.

47. PHGB, June 17, 1925. The December 1, 1922 U.S. Naval Policy was attached to the transcript of this hearing in 1925, the first year the 1922 policy saw significant revision.

48. PHGB, March 30, 1925, 190. Seniority and promotion date from GB ML. See also Miller, *War Plan Orange,* 135–136. Schofield is perhaps best remembered for his leader-

ship of the War Plans Division in the late 1920s and later became Commander-in-Chief of the U.S. Fleet (CINCUS).

49. The "blueprint" in use was a distinct document from the "serial" on the same topic. The "blueprint" for the naval policy was actually a policy that would be promulgated to the entire Navy once approved; the serial was only for the General Board, select offices, and the secretary and was an "internal" staff study.

50. *Conference on The Limitation of Armament, Senate Document No. 126,* (Washington, D.C., 1922). The Washington Naval Treaty was signed on February 6, 1922.

51. PHGB, March 3, 1922, 52.

52. Admiral McVay wanted it clearly understood that the Navy was wholeheartedly convinced of the "airplane carrier's" value. Some of this wording was no doubt a response to the actions of General Mitchell since the Navy air policy would receive wide distribution. The General Board wanted to avoid any occasion for criticism of its air policy.

53. PHGB, March 3, 1922, 56–57.

54. NARA RG80 GB, "Some Effects of the Washington Conference on American Naval Strategy," by Captain F. H. Schofield, lecture delivered at the Army War College, Washington Barracks, September 22, 1923.

55. For examples see, RG38, Joint Army and Navy Problem 1922–1923, (NWC–AN3/GSS–EF7) and Strategic Problem IV, February 9, 1926. These problems were also held in the OpNav files of the War Plans Division.

56. NARA RG80 GB 438-1, May 17, 1930 Telegram to senior member present General Board, signed Bradley A. Fiske.

57. NARA RG80 GB 438-1 Serial 1730, October 17, 1936. This serial was signed by the senior member present Adm. J. M. Reeves.

58. Miller, *War Plan Orange*, Chapters 6 and 7 passim.

59. Chief of Naval Operations March 4, 1922 memo to GB, passim; NARA RG80 GB 438-1, April 27, 1922, transcript of speech of Rear Adm. H. S. Knapp to the American Society of International Law. See also the discussions above regarding the criticism by naval leaders of the Treaty.

60. NARA RG80 GB 420-2, Chief of Naval Operations "Memorandum in Re: Tentative Draft of U.S. Naval Policy," March 4,1922, 2–7. This memo was attached to Serial 1120 and is representative of these views. It includes handwritten annotations by the Chief of Naval Operations Adm. R. E. Coontz, which emphasized that Guam and the Philippines must be maintained in the "best condition possible consistent with treaty limits."

Chapter 4

1. HL, Memo from the Secretary of the Navy for the General Board, November 12, 1931.

2. NARA RG80 GB 420-2 Serials No. 1022, 1055, and 1130. Serial 1130, May 31, 1922 discuss these issues and then lay out justification for the design "blueprint" for the fleet. The first post-Washington *U.S. Naval Policy* was promulgated in December 1922 (see Appendix 2).

3. NARA RG80 GB 420-2, March 2, 1922 letter from BuAer.

4. NARA RG80 GB 420-2 Serial 1130, 10. This serial reflects in its language the *U.S. Naval Policy* of 1922 but includes additional justification as well recommendations for the future. It appears the authors of this serial were Admiral Rodgers and Captain Schofield.

5. Ibid., 1–4.

6. Ibid., 1, 4–5. The Harding administration planned on hosting another conference in 1924. The details of battleship modernization were covered in G.B. Serial No. 1121. See Chapter 5.

7. Ibid., 4–6. See also Philip T. Rosen, "The Treaty Navy, 1919–1937," in *In Peace and War*, 2nd ed., K. J. Hagan, ed. (Westport, CT: Greenwood Press, 1984), 223.

8. C. Northcote Parkinson, *Parkinson's Law: The Pursuit of Progress* (London: John Murray, 1958). This rule is sometimes known as Parkinson's Law after C. Northcote Parkinson, a British naval historian. It was formulated, in part, to lampoon the Royal Navy.

9. PHGB, January 17, 1925, *U.S. Naval Policy* signed December 1, 1922 by Edwin Denby, Secretary of the Navy, attached to General Board hearing transcripts (see Appendix 2). "Armament" refers to gun caliber.

10. *U.S. Naval Policy* of 1922. Aircraft Carrier tonnage varies because the Navy had already expended 66,000 tons of its 135,000 tons of allowed carrier tonnage in the conversion of *Lexington* and *Saratoga*. This left 69,000 tons for additional carriers—so the Navy could build at most two 27,000 ton carriers. The first carrier size recommended by the General Board was 27,000 tons. See also NARA RG80 GB 449, November 18, 1922, "Second Endorsement to Naval Aeronautic Policy," 3.

11. NARA RG80 GB 420-2 "Memorandum from the General Board for File, Subject: Evolution of U.S. Naval Policy," December 5, 1922. This document catalogues the changes to the 1922 policy that were published as the 1923 policy. I believe that the General Board created this document to have a written version of the Secretary's oral instructions and their recommendations "on the record."

12. NARA RG80 438-1, December 10, 1923 letter from London Naval attaché forwarded to General Board refers to the machinations by the United States for another conference. The attaché claimed that the British wanted to include limits on aviation and lock in their superiority in cruiser tonnage.

13. It is worth noting, as a measure of the Harding administration's attitude toward the Navy, that the Teapot Dome Scandal revolved around the transfer of strategic Navy oil reserves and fields to the Interior Department and its secretary's fall, under whose leadership the scandal occurred.

14. NARA RG80 GB 438-1 Secret Serial No.1239, acting Secretary of the Navy letter, August 18, 1924. The European situation no doubt refers to the deep engagement of Europe's diplomats in the League of Nations Geneva Protocols that included proclamations on general arms reductions.

15. Ibid., 1.

16. PHGB, roll 1 index to General Board serials.

17. Braisted, "On the General Board of the Navy, Adm. Hilary P. Jones, and Naval Arms Limitation, 1921–1931," in *The Dwight D. Eisenhower Lectures in War and Peace, No. 5* (Manhattan, KS: Kansas State University, 1993).

18. NARA RG80 GB 438 Serial No. 1239, June 3, 1925, 1–3.

19. Ibid., 3. The secretary's response occurred almost a month after the General Board's finalized serial had been forwarded.

20. NARA RG80 GB 420-2 Serial No. 1338, December 11, 1926.

21. NARA RG80 GB 420-2 Serial No. 1345, April 5, 1927. This serial was signed by Admiral Jones.

22. Serial 1345. Reference (e) was Serial 1315 of the previous year.

23. RG80 Correspondence of the chief of naval operations 198-1 (microfilm), December 30,1923 Appendix F (originally classified secret) to WPL-9 War Plan Orange, Mobile Base Project. Hereafter referred to as 1923 Mobile Base Project.

24. NARA RG80 GB Serial No. 1162, April 7, 1923, "Naval Policy/Building Policy for 1926."

25. Serial 1345. See also Norman Friedman, *U.S. Aircraft Carriers: An Illustrated Design History* (Annapolis: Naval Institue Press, 1983), 67–71. Friedman gives great weight to the inputs of BuAer and the War Plans Division in getting the General Board to agree to the smaller size. Friedman's conclusions are somewhat at odds with the General Board record which indicates that up to April of 1927 the Board still very much wanted a 23,000-ton carrier. The General Board formally recommended a 13,800-ton carrier only in Serial 1376 of 1928. Subsequent tactical justifications from the three organizations—Board, OpNav, and NWC—read like rationalizations only made after the acceptance of the reality that they could not get more.

26. NARA RG80 GB 420-2, H.R. 8687, May 26, 1924. Attached to serial 1271.

27. NARA RG80 GB 420-2, September 27, 1926 letter from Admiral Pratt to the General Board. Admiral Pratt used the term "best seller" in referring to the General Board's new policy of calling its building program a "replacement" program instead "an increase in the navy." The idea was actually relayed to the Board by a liaison officer in Great Britain who adapted it from his Royal Navy counterparts who were also having similar budgetary challenges.

28. Serial 1345. This 1927 serial was the first to see the Navy reluctantly scale back the 27,000 tons in the hopes of getting the money for the smaller ship.

29. NARA RG80 GB 438-1, January 23, 1926. Letter from the General Board to Rear Admiral Pratt, president of the Naval War College. The "precept" was the Board's recommended guidance on the subject.

30. NARA RG80 GB 438-1 March 2, 1926 London Naval Attaché's report, 8. William F. Trimble, "Adm. Hilary P. Jones and the 1927 Geneva Naval Conference," *Military Affairs*, vol. 43, No.1 (Feb 1979), 2. See also Braisted, "On the General Board of the Navy."

31. NARA RG80 GB 438-1, March 2, 1926, London Naval Attaché's report, 4, 8.

32. Captain F. H.Schofield, "Some Effects of the Washington Conference on American Naval Strategy," 1–2, 5. Schofield first emphasized the importance of maintaining and

modernizing the battleships. His discussion of submarines emphasized their usefulness, especially the "cruiser submarine," in the absence of "outlying bases." Later, as War Plans Division Chief, Schofield testified similarly during the 1927 Board hearings on the desired characteristics for cruiser submarines.

33. Norman Friedman, *Submarine Development and Design* (London: Conway Maritime, 1984), 35.

34. PHGB, roll 1 Index and April 10, 1925, "Submersible Fuel Tender." This meeting, chaired by Jones and attended by Captain W. S. Pye, one of the authors of the original War Plan Orange, discussed a proposal by OpNav for using submersible fuel tenders to increase the range of the Navy's S-class submarines in the Pacific. Not only fuel, but also spare parts and personnel for augmentation, were to be carried. The General Board produced no less than seven serials on the topic of increasing submarine operational ranges in 1924–1925.

35. PHGB, April 10, 1925, "Submersible Fuel Tender," 219–220.

36. See Heinrichs, 199, for the view that starving bureaucracies are more innovative.

37. Trimble, 2–3.

38. Norman Friedman, *U.S. Cruisers* (Annapolis, MD: Naval Institute Press, 1984), pp. 112, 472.

39. Baer, 109, Buckley, 1993, 645, and Love, 554–555 all cite a fight without bases as the reason for the Navy's commitment to the 10,000-ton cruiser. The Japanese, meanwhile, were already building cruisers in excess of this figure.

40. Trimble, 2.

41. H.R. 8686, May 26, 1924 (copy attached to GB 420-2 Serial No. 1271).

42. Braisted, 1994, "On the General Board of the Navy."

43. Trimble, 2–3. Asada, "From Washington to London," 164–165.

44. Asada, 1994, 164–166. Trimble, 3–4. The Japanese worked out an initial agreement of a 65 percent ratio vis-à-vis the British for cruisers and destroyers. Eventually this figure was 63 percent of 500,000 tons or 315,000 tons, which would have given the Japanese a slight superiority in cruisers given the projected U.S. building program.

45. Braisted, passim. Trimble, 2–4.

46. Samuel E. Morison, *The Two-Ocean War* (Boston: Little, Brown & Co., 1963), 7. The original bill had called for five aircraft carriers and twenty-five new cruisers to be constructed in the next nine years. The carrier approved was the 13,500-ton *Ranger*, the first aircraft carrier to be authorized since the Washington Conference battlecruiser conversions.

47. NARA RG80 GB 420-2 Serial No. 1415, April 4,1929. The General Board used this serial to reiterate its "master plan" for the treaty navy to the Hoover administration. Included was verbiage that again emphasized that a floating dry dock would be "indispensable" in time of war. The Secretary of the Navy responded with a memo on April 20, 1929 (attached to the serial in the archives) that the serial of the previous year (1383) had been "promulgated," that is, accepted.

48. http://www.chinfo.navy.mil/navpalib/ships/battleships/bbhistory.html, accessed 05 July 2006. The division of the fleet occurred shortly after the Washington Conference in 1922.

49. Mahan, "The Battle of the Sea of Japan," 471.

50. NARA RG80 GB 438-1 Serial No. 1390, September 21, 1928. The General Board rejected this pact "even as a basis for discussion."

51. Braisted, passim. Jones' tables that led to the yardstick can be found in NARA RG80 438-1, February 28, 1928, a letter he wrote to the Secretary of the Navy which was also endorsed by the Chief of Naval Operations Adm. C. F. Hughes as the basis for further General Board discussion. NARA RG80 GB 438-1, September 24, 1929 memorandum, from Hoover to the General Board. This important memo lists five different forms of the yardstick formula: the General Board's, Admiral Jones', and three proposals by Hoover.

52. NARA RG80 GB 438-1 Serial No. 1444-A, September 11, 1929 memorandum. This memo was attached to the serial and shows the great lengths the new president went to in order to accommodate both the British and the General Board.

53. NARA RG80 GB 420-2, September 7, 1928, 1–13. This document was a lengthy memorandum by Admiral Moffett forwarded to the General Board by the Chief of Naval Operations in favor of accelerated aircraft carrier construction. In the memo, Moffett extensively cites recent secret operational after-action reports by the Commander of the Battle Fleet complaining that "Additional aircraft carriers are badly needed." Moffett used the memo as an occasion to give a short history of the neglect of carriers since the Washington Naval Treaty.

54. PHGB, November 27, 1929, 398.

55. PHGB, March 11, 1925, 98–104. This Op-12C War Plans Division study was attached to testimony given by Capt. M.G. McCook before the General Board.

56. *U.S. Naval Policy*, 1922, "Building and Maintenance Policy," (see Appendix 2).

57. Braisted. NARA RG80 GB 420-2, Pratt to Long, September 27, 1926. Pratt was under no delusions about the difficulty of getting new ships built. Pratt agreed with Admiral Long of the General Board to use the word "replacement" instead of "new construction" for all naval construction under the treaty would make it more palatable to politicians.

58. Richard Dean Burns, ed., *Encyclopedia of Arms Control and Disarmament, Volume IIII* (New York: Charles Scribner's Sons, 1993), 647.

59. Asada, 178–179. See also James B. Crowley, *Japan's Quest for Autonomy: National Security and Foreign Policy, 1930–1939* (Princeton, NJ: Princeton University Press, 1966), 66–71, for a retelling of the political crisis that occurred in Japan as a result of the London Naval Treaty.

60. Buckley, 647. Italy and France were allowed to construct 70,000 tons of new capital ship construction.

61. HL, Box 34, Cabinet Papers, November 22,1930 memo to the Secretary of the Navy. Hoover would go even further by delaying construction on all ships to generate good will for the 1932 Geneva General Disarmament conference. He closed many Navy bases and yards and for a time put the fleet on a 33 percent reduced operations schedule. See also NARA RG80 Correspondence of the Secretary of the Navy, September 7, 1930, "Reduction in expenditures in forces afloat."

62. NARA RG80 GB 420-2, BuAer May 28, 1930. This memo from Moffett emphasizes all these points and recommends construction of all flying deck cruisers allowable to meet the shortage of aviation in the fleet. See also Burns, 1177. From Article 3 and Article

16, paragraph 5 of the London Naval Treaty; Alan D. Zimm, "The U.S.N.'s Flight Deck Cruiser," *Warship International* No. 3 (1979), 220–221.

63. Buckley, "The Washington Naval Limitation System," 646–647. Goldman, 255–256. Asada, 183–184.

64. Baer, 116–117. Roskill, *Naval Policy Between the Wars, Vol. II*, 65.

65. HL, Box 38, Cabinet Papers, March 25, 1929 BuAer memorandum to Chief of Naval Operations. Moffett argued strenuously against this initiative. See also HL, Box 39, Cabinet Papers, May 22,1931, Treasury Department memorandum.

66. HL, Box 34, Cabinet Papers, Department of the Navy, October 24, 1930 was the date the memo was filed. The memo proposed that s a "special board" to be convened. The Chief of Naval Operations and the General Board were specifically rejected as forums for a fresh look at this problem. This memo was certainly written earlier since the filing date is after the October 16, 1930 General Board Serial (1475) that implemented the London Treaty for the Navy's building policy.

67. NARA RG80 GB 420-2 Serial No. 1475, October 16,1930, "Building Program—Fiscal Year 1932"; Norman Friedman, *U.S. Battleships: An Illustrated Design History* (Annapolis, MD: Naval Institute Press, 1985), 202–207. Modernization continued for the battleships throughout the period once the initial round of treaty-inspired changes were complete.

68. NARA RG80 GB 438-1, December 5,1931. This memo discusses the proposal, accepted by the United States, not to increase armaments for a year from November 1931. The General Board felt that this agreement need not prevent the construction of already approved ships. However, GB 438-1 of November 28, 1932 makes it clear that the Hoover administration interpreted the agreement to prevent "new" construction even if already authorized prior to the agreement. To make things worse the Hoover administration extended the building "truce" another four months. Hoover laid down no new ships, as a result of this last extension, during his entire term.

69. NARA RG80 GB 438-1, May 21,1931, 8. "Displacement and Gun Caliber of Battleships" by Cdr. E. M. Williams prepared for the executive committee of the General Board.

70. GB 438-1, May 21,1931, "Displacement and Gun Caliber of Battleships."

71. James Barros, "The League of Nations and Disarmament," in Richard Dean Burns, ed., *Encyclopedia of Arms Control and Disarmament, Vol. II*, 618–619.

72. Wayne S. Cole, "The Role of the United States Congress and Political Parties," in Dorothy Borg and Sumpei Okamoto, ed., *Pearl Harbor as History: Japanese-American Relations, 1931–1941* (New York: Columbia University Press, 1973), 304.

73. NARA RG80 GB 438-1 Serial No. 1521aa, January 18, 1933. This serial, entitled "Limitation and Reduction of Armaments" was renumbered as serial 1584.

74. http://www.presidency.ucsb.edu/ws/index.php?pid=14655&st=Annapolis&st1= (May 1, 2006); David Hoffman, film director, *Wings Over Water* (Camden, ME: Varied Directions, Inc., 1986). The transcript for FDR's June 1933 commencement address closed with the words "Keep the faith. Good luck in the days to come!" In the film FDR continues, "Let me tell you from the bottom of my heart as one who can only say unofficially to you,

but who does say unofficially to you, that he loves the United States Navy more than any other branch of our government."

75. http://www.historicaldocuments.com/NationalIndustrialRecoveryAct.htm, April 25, 2006. Quoted from section 202 of the Act passed June 16,1933, "and if in the opinion of the president it seems desirable, the construction of naval vessels within the terms and/or limits established by the London Naval Treaty of 1930 and of aircraft required therefore. . . . "

76. Baer, 128–130. Buckley, 649–650.

77. Norman A. Graebner, "Hoover, Roosevelt, and the Japanese," in Dorothy Borg and Sumpei Okamoto, ed., *Pearl Harbor as History: Japanese-American Relations, 1931–1941* (New York: Columbia University Press, 1973), 34–35; Asada, "The Japanese Navy and the United States," 240–243. Asada claims American actions played a role, but that the trend was toward the abandonment of the treaty system anyway principally due to the efforts of the Adm. Kato Kanji. Kato was the head of the so-called "fleet faction" that had opposed the Washington Treaty from its inception.

78. Friedman, *U.S. Battleships*, 243.

79. Ibid., 270–274. The Japanese were already planning to build the 18-inch-gun "super-battleships" *Yamato* and *Musashi*.

80. For example see Lt. Cdr. Melvin F. Talbot, "The Battleship: Her Evolution and Her Present Place in the Scheme of Naval War," 1938 U.S. Naval Institute Prize–winning essay, in *United States Naval Institute Proceedings*, Vol. No. 64, No, 5 (May 1938), 645–653. Talbot's essay is a typical defense of the battleship of the period—"She, and she alone, remains an absolute." A similar essay had won the prize essay in 1926. The year of the Washington Conference was the only year during the interwar period when a prize essay was not awarded.

81. PHGB, March 27, 1919, "Development of Naval Aviation Policy." This was the famous hearing where the young John Towers predicted that the "airplane carrier will not last very long" probably due to improvements to the battleship as naval technology advanced.

82. The view of the U.S. Navy as a balanced fleet at the end of the 1930s is still a matter of debate. See Thomas C. and Trent Hone, *Battle Line: The United States Navy, 1919–1939* (Annapolis, MD: Naval Institute Press, 2006), 1. The Hones argue instead that the Navy was three distinct fleets. However, three distinct fleets seem to imply, by the very fact they are distinct, some measure of balance.

83. H.P. Willmott, *The Battle of Leyte Gulf*, 3. See also Hammond, *The Treaty Navy*, 174. Both Hammond and Willmott agree that the U.S. Navy built during the interwar period fought the Japanese Navy to a standstill.

Chapter 5

Note: The epigraphs were taken from NARA RG80 GB 449, Joint Board August 18, 1921, "Report on Results of Aviation and Ordnance Tests." This report was included with GB Serial 1110 and attached information on the bombing of the *Ostfriesland*. It was signed by General

Pershing, Acting Secretary of the Navy T. Roosevelt Jr., and the Secretary of War John Weeks. It was a stinging rebuke to the exaggerated claims of General Mitchell; Melvin F. Talbot, "The Battleship: Her Evolution and Her Present Place in the Scheme of Naval War," Prize Essay 1938, *United States Naval Institute Proceedings*, vol. no. 64, no. 5 (May 1938), 653.

1. 1924 War Plan Orange, WPL-9, passim.

2. *U.S. Naval Policy* 1922, Building Policy Section (see Appendix 2). Modernization was the interwar Navy's characterization of its program of improvements for the eighteen retained battleships under the Treaty. This characterization did not change during the period, see also GB 420 Serial 1383, June 11, 1928, the language was verbatim.

3. See Appendix 1, chapter 2, part 3, paragraph d. Most authors refer to this clause as the "reconstruction clause," as does this study. Others, notably Norman Friedman, refer to it as the "modernization clause." See Friedman, *U.S. Battleships*, 189–191.

4. See Appendix 1 for a list of these retained ships. The two oldest battleships, *North Dakota* and *Delaware*, were to be scrapped as soon as construction was complete on the two newest ships *West Virginia* and *Colorado*.

5. Battleships often had two or more "batteries" of guns. The main battery consisted of the largest caliber weapons, often from nine to twelve guns housed in multiple turrets. The secondary battery was used at shorter ranges against enemy cruisers and destroyers, and so on.. A third battery at this time consisted of weapons designed for air protection or to repel boarders. More and more the secondary battery came to have a dual role in both surface and air defense.

6. PHGB, April 17, 1922, "Interpretation of Treaty RE Modernizing Capital Ships." Pratt was also a member of the General Board, but testified in his capacity as a recent Washington Conference adviser. The detail of modernization allowed under the treaty was contained in the so-called "reconstruction clause," which can be found in Article XX, Chapter II, Part 3, paragraph (d) (see Appendix I).

7. See Samuel Huntington, *The Soldier and the State* (Cambridge, MA: Harvard University Press, 1957), Chapter 4 passim. Huntington's classic remains the classic argument in support of this point: that modern militaries are by their very nature conservative.

8. Trent Hone, "Evolution of Fleet Tactical Doctrine in the U.S. Navy, 1922–1941," *Journal of Military History*. vol. 67, no. 4 (October 2003), 1117–1118. Hone's essay is the accepted interpretation of interwar battle line tactics. See also Friedman, *U.S. Battleships*, 156–157. PHGB, February 21, 1921, "Characteristics of Airplane Carriers," discusses the General Board's desire to convert some of the 1916 plan battle cruisers to carriers given their speed and modern propulsion.

9. NARA RG80 GB 449, July 29, 1921 secret report from Commander Van Keuren to the Chief of the Bureau of Construction and Repair. William D. O'Neil, "Transformation, Billy Mitchell Style," *United States Naval Institute Proceedings*, March 2002, 100–104.

10. NARA RG80 GB 438-1, January 6, 1922, "Minutes of the Tenth Meeting of the Advisory Committee," 7.

11. Ibid., 11.

12. Buckley, "The Washington Naval Limitation System," 642. This remark is attributed to Charles A'Court.

13. See Appendix 1.
14. See Appendix 1, Chapter 2, Part 3.
15. GB 420-2, Serial No. 1121, attached Joint Letter April 4,1922 from Bureaus of Engineering, Aeronautics, Ordnance, and Construction and Repair.
16. Ibid.
17. Friedman, *U.S. Battleships*, 436–437. Friedman cites performance for the *Texas* as 7,600 nm at 12 kts prior to the conversion and 15,400 nm at 10 kts after the conversion to oil-burning.
18. GB 420-2, March 2, 1922 memorandum from BuAer.
19. PHGB, April 17, 1922, "Interpretation of Treaty RE Modernizing Capital Ships." This was the same Captain Schofield whom Admiral Rodgers appointed to lead the discussion on the *Naval Policy*. Rodgers was also present on this occasion as the senior member.
20. PHGB, April 17, 1922, 373 and passim.
21. PHGB, April 18, 1922, "Modernization of Existing Capital Ships," 391 and passim. The five newest battleships were the *Tennessee, Maryland, California*, and those allowed for completion under the Treaty the *Colorado* and *West Virginia*. These five were know as "The Big Five" and were the last to be modernized. See also Friedman, *U.S. Battleships*, 203.
22. GB 438-1 Serial No. 1121, April 26,1922.
23. GB 420-2 Serial No. 1130. See also Friedman, 190. Friedman believes that the Board did not include electric drive because of its worry over the cost of the program in the current "anti-naval atmosphere." Certainly this played a role, but the hearings make it clear that the Board thought the idea of using the extra electric drives a good one but there was concern over their legality under the treaty. This was in spite of the fact that the Board had adopted the principal that the Treaty did not prevent internal modifications as long as weight limits were not exceeded by the extra 3,000 tons allowed by the reconstruction clause.
24. Friedman, 1985, 193 and 432–437.
25. PHGB, April 18, 1922. See also Friedman, Chapter 10. The seven older battleships were the *Nevada, Arizona, Oklahoma, Pennsylvania, New Mexico, Idaho*, and *Mississippi*.
26. GB 449, November 8, 1921. Admiral Taylor of the Bureau of Construction and Repair had in fact made this very recommendation in a letter to the General Board before the Washington Conference had even convened. "If the opportunity should be had to carry out such experiments on a fairly modern ships, this . . . would be of great value." Perhaps Rodgers, who was both on the Board and a delegate at the Conference, had influenced the drafting of Article XX with this very goal in mind. Article XX, Chapter II, Part II (d) pertains. See Appendix 1.
27. GB 438-1, April 22, 1922 letter from Admiral Rodgers to Colonel Lucas.
28. H.B. Grow, "Bombing Tests on the 'Virginia' and 'New Jersey,'" *U.S. Naval Institute Proceedings* vol. 49, no. 12 (December 1923). Grow emphasized that the battleship remained "the first line" but that the "officers in the fleet aid the progress of naval aviation in every way possible."
29. Friedman, 1985, 186.
30. GB 438-1, November 17, 1924 letter from the Secretary of the Navy to the president. Presidential approval was attached and dated with the same date.

31. Admiral William S. Sims was a key exception, deciding after Naval War College gaming that carriers were the capital ship of the future. See Richard Hough, *Death of the Battleship* (New York: Macmillan Co., 1963), 22–23.

32. Alan D. Zimm, "The U.S.N.'s Flight Deck Cruiser," *Warship International* No.3 (1979), 231. Zimm argues that this dedication, at least in 1930, was perfectly understandable given the defensive power of battleships versus the capabilities of the aircraft of that day.

33. Trent Hone, 1108, 1117–1118. The British force at Jutland included a very fast division of new battleships that allowed them to out steam the German battle line and "cross the T," that is concentrate converging fire on the head of the enemy column.

34. The London Naval Treaty of 1930 later modified these further, limiting new construction battleships to 14-inch guns and extending the building holiday.

35. U.S. Department of State, *Proceedings of the London Naval Conference of 1930 and Supplementary Documents*, Conference Series No. 6 (Washington, D.C.: U.S. Government Printing Office, 1931), 82. None of these proposals got off the ground, although they were seriously proposed. For example, Secretary of State Stimson proposed to abolish the submarine at London in 1930, but the objections of France and Japan scuttled this proposal.

36. RG-38, *Basic Orange Plan*, February 1924, 73. Below waterline damage, in particular, is almost impossible to repair without a dry dock. The Navy could make battle repairs by sealing off compartments and then using lumber to shore up (buttress) adjoining passageways, but these sorts of repairs are for survival and not permanent.

37. Nils Gilman believes that Americans in particular are susceptible to this "two sides of the coin" sort of thinking, being optimistic on the one hand while anxious on the other. Naval officers were no exception, but as the period wore on they became less and less optimistic, even the normally optimistic and progressive Admiral Pratt. Nils Gilman, *Mandarins of the Future* (Baltimore, MD: Johns Hopkins University Press, 2003), ix–xi, 270–273.

38. Friedman, *U.S. Battleships*, 118. The General Board had considered the 16-inch gun caliber as early as 1911. By 1914 BuOrd testing of the new 16-inch design had proved its superiority to the existing 14-inch guns in explosive power and range.

39. Trent Hone, 1113. These superior ships were the *Colorado* class. The U.S. Navy had three of them to two for Japan and Great Britain. Friedman, 183–186. Friedman in part blames U.S. Naval intelligence for not properly assessing the speed of other nations' battleships, which contributed to the slower speed of most of the U.S. battle line.

40. Trent Hone, 1113. Twelve the of the eighteen U.S. battleships retained under the Treaty were armored in this way. The others could not be modified due to the Treaty language preventing "alterations in side armor."

41. Ibid., 1115–1117. Friedman, *U.S. Battleships,* 172–173. The definition for RMA implied here is that found in Murray and Knox, 12. They define an RMA as tactics, technology, or organization that provides "new ways of destroying opponents."

42. Friedman, *U.S. Battleships*, 85. The kite balloon technique used a modified balloon attached to the ship to provide extended line of sight for the spotting of gun fire. It was limited by the optics used by the observer, intervening weather, and maneuvering and speed of the ship.

43. PHGB, March 8, 1919, "Developments in Aviation." The transcript annotation was probably approved by the head of the General Board Adm. Charles Badger and at the very least by Winterhalter, whose words they were.

44. GB 420-2, March 2,1922 memorandum from BuAer. This recommendation was included among those Moffett drafted in response to the Board's request for inputs for the *Naval Policy.*

45. PHGB, April 17–18, 1922.

46. PHGB, December 5,1922, "Modernization of Existing Capital Ships." Admiral Huse conducted this hearing and opened it by citing a Bureau of Ordnance report of October 26,1922. From this report he pointed to the grave difference between what the "majority" of the British fleet could range versus the "majority" of the U.S. fleet—30,000 yards versus 21,000 yards. It's clear Huse's prority for modernization of turrets were the older coal-fired battleships.

47. Ibid. The issue of turret modification had not been previously broached with respect to the Washington Treaty on April 17 of that year, when the Board had discussed at length what it thought was allowed under the treaty for battleship reconstruction. See Appendix I, Article XX, Chapter II, Part 3 (d).

48. PHGB, December 5, 1922, 817–819.

49. Ibid., 831. Pratt was also present but said nothing on the record on the issue of the turrets. Friedman, *U.S. Battleships*, 82–85. Friedman dates the Board's authority over ship design from after the Newport Conference of 1908, when the General Board took over the role of initiating and ship characteristics and concepts from the Board of Construction, which was dissolved.

50. PHGB, December 5, 1922, 832. GB 420-2 Serial No. 1155, December 19, 1922. The serial recommended that "(1) That angle of elevation of all turret guns of thirteen ships antedating the *Tennessee* class be increased to thirty degrees and that sufficient money be made available immediately for this purpose. Handwritten next to this on the serial for the Secretary of the Navy was 6,500,000 dollars.

51. NARA RG38 Correspondence of the Secretary of the Navy, January 5, 1923.

52. Friedman, 190–191. GB 420-2 Serial No. 1162, April 7, 1923.

53. GB 438-1 Serial No. 1195, November 1, 1923.

54. GB 420-2 Serial No. 1206, April 18, 1924. Jones argued "The alterations necessary to give [battleships] these characteristics [turret elevation alterations] are perfectly within our treaty rights but funds are needed to effect them."

55. H.R. 8687, attached to GB 420-2 Serial 1206. This bill approved "installation of additional protection against submarine attack, of installation of antiair [sic] attack deck protection, of conversion of such vessels to oil burning, and, in addition, for the *New York* and *Texas*, the purchase, manufacture, and installation of new fire-control systems."

56. GB 420-2, February 16,1925 memorandum from President Coolidge to the General Board. Admiral Jones had argued the previous November for elevation modifications for all thirteen older battleships and made the same arguments again in his next serial in April 1925. See GB Serials 1251, November 3, 1924 and 1271, April 3, 1925. The 1930

London Naval Treaty would allow the U.S. Navy to convert some of these ships into target vessels, for example *Utah*, which the Japanese sank at Pearl Harbor.

57. GB 420-2 Serial No. 1271, April 3, 1925; Serial No. 1313, March 30, 1926; GB 438-1, 27 June 1925 notes by Admiral Pratt of June 24, 1925 meeting with Mr. Hughes, Lake George, NY; Friedman, 197.

58. GB 438-1, January 18, 1926 Secretary of the Navy letter to Ambassador Houghton.

59. GB 449, July 31,1928 BuAer memorandum to the Secretary of the Navy. The report cited was Commander-in-Chief, U.S. Battle Fleet, July 1,1927 pages 6 and 18.

60. PHGB, December 5, 1930, 33. This remark was made by Admiral Bristol, former CINCAF, who was intimately familiar with the impact of the treaty system on the strategic situation in the Far East.

61. NARA RG38 War Plans Division, Naval War College Problems Series IIA, Strategic Problem IV, February 9, 1926, 4. These problems were provided to the College by OpNav.

62. GB 438-1, May 21, 1931, 8. "Displacement and Gun Caliber of Battleships" by Commander E. M. Williams prepared for the executive committee of the General Board.

63. GB 438-1, March 5, 1934 memorandum for the Board by TSW, "Reduction of size and gun power of battleships." TSW was in all likelihood Cdr. T. S. Wilkerson, the Board Secretary and later an admiral during World War II.

64. George T. Davis, *A Navy Second to None* (New York: Harcourt, Brace, & Co., 1940), *xii–xiii*.

65. Ronald Spector, *Eagle Against the Sun* (New York: Vintage Books, 1985), 66–74. In 1941 U.S. strategy as articulated by the Chief of Naval Operations Admiral Stark was to go on the strategic defensive in the Pacific, a stance that favored the "step-by-step" course of action. However, General MacArthur's dissatisfaction with this strategy led to possibility that the U.S. Fleet might execute the "thrusting" course of action in somewhat modified form if MacArthur was successful in holding the Philippines. Any possibility that a decisive naval battle would be sought early was eliminated by the crippling attack on the U.S. battle line at Pearl Harbor. Spector argues that Stark was never in favor of this course of action in any case, although he allowed Admiral Kimmel to plan for it.

66. *North Dakota* and *Washington* were used in extremis, at night, during the Guadalcanal campaign against Japanese battleships. This was only because the Pacific Fleet had been reduced to only one operational aircraft carrier and the situation at Guadalcanal had become critical.

Chapter 6

Note: The epigraph was taken from Alan D. Zimm, "The U.S.N.'s Flight Deck Cruiser," *Warship International* No.3 (1979), 221, excerpt from Moffett's diary.

1. PHGB, February 21, 1921, "Characteristics of Airplane Carriers," CDR John Towers stated BuAer policy that gun-spotting for battleships was the primary mission for carrier aircraft; NARA RG38, chiefl of naval operations War Plans Division Records, May 1927 Operations Problem II-1927, 11.

2. The Morrow Board Report, November 30, 1925, attached to G.B. 449 material with Serial No. 1250. The Morrow Board was a panel of senior military officers and civilians (e.g., General Pershing) that met to consider aviation "unification." One of its findings was to maintain naval aviation as a separate entity under the Navy and not under a new air ministry or service.

3. WPL, February 9, 1924, Basic Orange Plan, 32. See also Miller, Chapter 8, passim.

4. *U.S. Naval Policy of 1922,* Air Policy Section (see Appendix 2); GB 438–122 September 1923, Captain Schofield speech, 7–8; Chief of the Bureau of Aeronautics Memo, "New War Plans," April 12,1923, attached to GB Serial No. 1140.

5. Miller, 56–57, 140–141. Miller argues that the "quick relief" thrusting course of action lost its adherents around 1925 and that a "marginally saner policy" was adopted that recognized that Manila might not hold. However, Navy planners wanted to establish an expeditionary base in Mindanao while at the same time seizing an intermediate base in the Caroline or Marshall Islands. Schofield oversaw the planning for this new course of action when he became head of the War Plans Division in 1926. RG80, Secretary of the Navy Correspondence, CinC Asiatic Fleet (CINCAF) August 23,1924. The CINCAF correspondence notes Japanese protests re: Treaty violations in late 1923.

6. RG38, Records of the War Plans Division, Box 8, February 1924 Basic Orange Plan (hereafter 1924 Orange Plan), 13; RG38, War Plans Division, Strategic Problem IV, 1926, 2. Strategic Problem IV gamed a particular problem associated with the current Orange plan.

7. *U.S. Naval Policy of 1922,* Building Policy Section.

8. *U.S. Naval Policy of 1922,* Air Policy Section. This section states, "To determine by trial the practicability and desirability of operating planes from destroyers and submarines."

9. Hone et al., 160; Williamson Murray and Barry Watts, "Military Innovation in Peacetime," in W. Murray and A. R. Millet, eds., *Military Innovation in the Interwar Period* (Cambridge: Cambridge University Press, 1996), 394. Admiral Moffett was killed in 1933 during the crash of the dirigible airship *Akron*. Without Moffett's support the lighter-than-air program was scaled back, but the Navy did not abandon the idea completely until late into World War II.

10. GB 438, copy of May 1,1933 "Memo of Admiral Hilary Jones for Navy League," PHGB, February 15, 1934, "Flying Deck Cruiser." Admiral Clark was in the minority in this view. Admirals opposed to the flying deck vessel but supportive of aircraft carriers included Admirals Leahy, Yarnell, Taylor, Standley, and Schofield.

11. August 23, 1924, Commander in Chief Asiatic Fleet, attached to GB 449 Serial No. 1250, "Air Squadrons, Asiatic—Development of." These six aircraft were the first of a total of eighteen for the full squadron. This correspondence and others led to an extensive General Board hearing (December 10, 1924) on the topic for the design or conversion of more suitable tenders to replace the *Ajax*.

12. PHGB, March 3, 1922; 1922 *U.S. Naval Policy*; GB 420–2 Serial 1055 (Naval Building Policy for 1923).

13. PHGB, December 10, 1924, "Proposed Characteristics for Airplane Tenders." Evidently there was no common usage: the terms airplane tender, aircraft tender, and air depot ship were used interchangeably during the hearing. A War Plans memorandum was read

into the record by Captain Johnson who was BuAer's representative at the meeting. BuAer based its building requests on the current Orange plan, WPL-9 February 1924.

14. GB Serial No. 1250, November 21, 1924, the BuAer October 29,1924 memorandum, "Tender for the Asiatic Fleet," attached to the Board's serial; PHGB, December 10, 1924. *Wright* was the aircraft tender and Flagship of the "Fleet Base Force" (see Chapter 7).

15. PHGB, December 10, 1924.

16.. Ibid. *Langley* was converted from the collier *Jupiter.* GB 420-2 Serial 1162, the building plan for 1923 shows funds to build eighteen planes for the Asiatic tender, however, it reflects no funds for conversion or new construction.

17. Naval Historical Center, *Dictionary of American Naval Fighting Ships,* 8 vols., James L. Mooney, ed. (Washington, D.C.: Naval Historical Center, 1991), 1:100-101 and 3:507. Hereafter, NHC DANFS.

18. "Commander" is a term of pay grade and rank equivalent to that of a Lieutenant Colonel in the Army. The Navy also used the term in its normal sense as a commanding officer, for example Commander, Asiatic Fleet or Commander in Chief, U.S. Fleet.

19. Seaplanes required three levels of maintenance; daily, intermediate, and depot level. Depot level was the most extensive, usually requiring rebuilding or reinstalling new engines. Seaplanes, being part boat and part airplane, also required hull maintenance, to include scraping the bottoms clean of marine growth and corrosion work. A similar set of problems occurred with the submarine fleet. These vessels were habitable during the dry season in the Philippines but during the summer monsoons they often relocated to a port in China, such as Chefoo. As the political troubles in China increased it became harder and harder to berth in the roadsteads of the Chinese ports. The only solution was to increase the number submarine tenders for the Asiatic Fleet, another competing priority for the limited construction dollars. NARA RG80, Office of the chief of naval operations, Record Message General (RMG) April 18,1932 from CINCAF to OPNAV.

20. RG80 Formerly Classified Correspondence of the Secretary of the Navy (hereafter SECNAV CC), March 30,1928 Commander Aircraft Squadrons, Asiatic Fleet (CAS, AF) to Commander-in-Chief, Asiatic Fleet (CINCAF); RG80 SECNAV CC, April 24, 1928 CINCAF to BuAer.

21. SECNAV CC, August 23,1928 CAS, AF to BuAer via CINCAF.

22. SECNAV CC, September 11, 1928, CINCAF to BuAer.

23. SECNAV CC, November 20, 1928 BuAer to Chief of Naval Operations.

24. SECNAV CC, November 20,1928 BuAer to Chief of Naval Operations; NHC DANFS, 4:47. Langley

25. SECNAV CC, June 3, 1930 CINCAF to Chief of Naval Operations. NHC DANFS, 3:507.

26. GB 420-2 Serial 1629, October 13, 1933. Because of the London Naval Treaty the treaty-limited classes now included submarines, cruisers, and destroyers.

27. NHC DANFS, 2:220; Oral Interview with LCDR Archie Mills, USN (ret), May 27, 2006. *Curtiss* (AV-4), commissioned in 1940 and having been approved during the 1937 budget, was the first purpose-built tender.

28. GB 420-2 Serial 1629, October 13, 1933. HL, OpNav War Plans Memo, March 27,1929, from Admiral Schofield to Navy secretary. Schofield was another voice that was arguing against a more diversified approach, arguing that that the Navy should not waste its money building even small aircraft carriers but continue to build the biggest ones possible in order to put more aviation forward with a fleet that would lack land-based air support in the western Pacific.

29. HL, Cabinet Office Papers, Box 34, April 19,1929 memorandum from H. M. Lord. Lord recommended that one large carrier be placed in a "reserve commissioned status" that June. 22 November 1930 memorandum from the president.

30. See Appendix 1. See Chapter 5 for discussion of the reconstruction clause and Appendix 1 (Chapter II, Part III).

31. PHGB, February 21, 1921, "Characteristics of Airplane Carriers"; PHGB, February 10, 1922, "Proposed Characteristics of Airplane Carriers converted from Battle Cruisers," by the time of this hearing the 35,000 tons had grown to 38,900 tons listed as the "former normal displacement." Friedman, *U.S. Aircraft Carriers*, 43; PHGB, June 8,1923, "Characteristics for Airplane Carriers." At this hearing Admiral Strauss made it clear that the Board's policy was that only 66,000 tons from these ships counted toward the total of 135,000 allotted to the United States for aircraft carriers, leaving 69,000 tons for any new construction.

32. George Wilson, February 17,1922 letter attached to GB 438-1 Serial No. 1088-1.

33. Hone et al., 47. *Saratoga* and *Lexington* would take twice as long to complete and join the fleet as originally projected, barely making it to the fleet for the battle problems of 1929 (Fleet Battle Problem IX).

34. GB 420-2 Serial No. 1251, November 3, 1924; see also SECNAV CC, Office of Naval Intelligence transcript October 23, 1923 meeting of naval officers with Westinghouse Company representatives, testimony of LCDR Grow, 18.

35. GB Serials 1239 and 1251. Representative Burton French, November 20, 1924 letter to the Secretary of the Navy attached to Serial 1251. French may have had an alternative motive for wanting smaller carriers, he would later propose to lay up the large carriers once a small one was built in order to save money.

36. Calvin Coolidge, February 16, 1925 memorandum to the General Board, attached to serial 1251. This was an unusual correspondence in that it was signed by the president and provided directly to the General Board, not the Navy secretary.

37. The Teapot Dome Scandal had resulted in the departure of Secretary Denby with T. R. Roosevelt Jr. serving acting as secretary until after the 1924 election. It had involved oil leases for Navy fuel reserves.

38. Friedman, 44, refers to Wilbur as "legalistic." Wheeler, *Admiral William Veazie Pratt*, 235–238; PHGB, May 27, 1925, "Reduction in Tonnage of Aircraft Carriers," Wilbur's May 26, 1925 letter to the Board was transcribed into the hearings for that day.

39. Wheeler, 236; Friedman, 44; GB 438-1, May 23,1925 memorandum for the Secretary of the Navy.

40. William Pratt, "Notes on the Interview Held with Mr. Hughes at his Home on Lake George, New York, June 24, 1925," attached to GB memo of May 25, 1925 for the Secretary

of the Navy. This correspondence was located with GB serials 1239 and 1251. See also PHGB roll 1, June 27, 1925 minutes and Wheeler, 236.

41. Miller, 147–149. Miller particularly highlights the problems posed to Orange by not building the key floating dry docks for the mobile base force as well as Schofield's frustration with the maintenance capability these docks would provide even if built.

42. Zimm, 220; Till 221; Hone et al., 48–49. Fleet Problem IX was the debut of the *Lexington* and *Saratoga*. It included Pratt and King as commanders of the forces and highlighted the offensive potential of the carriers, but it also highlighted their vulnerability.

43. PHGB, February 15, 1934, "Flying Deck Cruiser." The amount cited by 1934 was 19–20 million dollars.

44. Miller, 177.

45. BuAer July 31, 1928 to Secretary of the Navy, attached to GB Serial No. 1345. Moffett wrote this extremely important thirteen-page memorandum on the occasion of the development of the 1929 aircraft building policy.

46. BuAer July 31, 1928. Pratt's endorsement was attached to this correspondence. Moffett included CINCUS' annual report for 1927 [A9-1/OF1(4)] ending June 30, 1927 in his memorandum as well as Pratt's endorsement.

47. Cited in BuAer July 21, 1928.

48. Op-12-CD, October 8, 1928, memorandum from War Plans Division to Chief of Naval Operations and BuAer. This correspondence was attached to Serial 1376, the building program for 1930.

49. BuAer July 21, 1928, 8.

50. Ibid., 8–9.

51. David Evans and Mark Peattie, *Kaigun* (Annapolis: Naval Institute Press, 1997), 244. These ships used float planes and catapults and played a critical role at Midway in 1942. They were probably a response to the U.S. flying deck program.

52. PHGB, June 8, 1923.

53. PHGB, March 11, 1925, "Characteristics of Light Cruisers." The correspondence of the office of chief of naval operations, March 2, 1925 memorandum, "Light Cruisers—Type best suited to the needs of the United States" was read into the transcripts for this hearing.

54. GB 438-1 Serial No. 1464, January 3,1930; BuAer November 23,1929 to Secretary of the Navy, forwarded December 3, 1929 to the General Board.

55. Burns, 1177 and 1181. Zimm, 221, discusses Moffett's alleged subterfuge in proposing the flying deck cruiser at London.

56. See Roskill, vol. II, 81–82, and Watts and Murray, 394, for the most critical view; for a more moderate stance see Hone et al., 191; for the most favorable view of the flying deck cruiser see Zimm, 243–245.1 Zimm posits that the concept wasn't given a proper chance to be tested and was cancelled "without a trial."

57. Zimm, 221.

58. Ibid., 216.

59. PHGB, April 9, 1930, "Light Cruisers NOS 37 to 41 Preliminary Design." See also GB 420-2 Serial No. 1492, June 14,1930 and BuAer 28 May 1930 to the Secretary of the Navy. Moffett's memo is attached to Serial 1492.

60. GB Serial 1507, November 13, 1930. Pratt criticized the use of the word "hybrid" in his endorsement dated November 12 attached to the Board's serial.

61. Serial 1507, November 13,1930. Laning's discussions with Bristol and the executive committee are reflected in the PHGB, November 5, 1935 minutes.

62. PHGB, November 25, 1930, "Characteristics of 6-inch cruisers (No Landing-on Platform)." Moffett was also at this meeting. Zimm, 242, characterizes Laning as "a brilliant innovator."

63. PHGB, December 4, 1930, "Military Characteristics of Cruisers with Landing—on Decks." This was the first hearing and included the Chief of Naval Operations (Pratt), Bureau Chiefs of Aeronautics and Navigation, Admiral Taylor of War Plans, Admiral Clark of Chief of Naval Operations Training Division, senior representatives from all the other Bureaus, and the Assistant Secretary of the Navy for Air (Ingalls).

64. Ibid. Just prior to World War II *Lexington* had her heavy 8-inch turrets removed as did *Saratoga* in early 1942, Friedman, 51. Up to that point the heavy operational schedules of these ships had precluded the necessary yard periods to modify them.

65. Friedman, 58.

66. PHGB, December 4, 1930.

67. PHGB, December 4,1930, "Military Characteristics of Cruisers with Landing—on Decks."

68. Ibid.

69. Ibid.

70. Ibid.

71. PHGB, December 5, 1930. Flush deck refers to the absence of any structures above the flight deck as on *Langley*.

72. Ibid.

73. Ibid., including Appendix A, a memorandum for the General Board from Admiral Clark. Clark opened the letter with the statement: "From the trend of testimony by aviation personnel given before the General Board it appears that attacks from the air upon our surface ships are beyond the control of our air force."

74. NARA RG38 Correspondence of the chief of naval operations, December 18, 1930 from Chief of Naval Operations to president of the Naval War College.

75. PHGB, December 23, 1930, "Military Characteristics of Cruisers with Landing—on Decks." Admiral Yarnell was absent. BuC&R December 26, 1930 "Tabulation of Aviation Facilities afforded by Designs of Ten Thousand-ton Cruisers with Landing—on Deck submitted to the General Board by the Bureau of Construction and Repair," attached to December 23 transcript.

76. Ibid.

77. PHGB, minutes from December 26 and 30, 1930, and January 23, 1931. Zimm, 233. Zimm cites a 03 February 1931 Chief of Naval Operations letter to the bureaus that contained "tentative characteristics."

78. Zimm, 221–225. Zimm uses the Naval War College report sent to the General Board after the games were over. PHGB, December 5, 1930, December 6, 1930 OpNav memorandum for the General Board, "Bombing on Light Cruiser Targets," attached as

Appendix B to the hearing transcript. Willis Lee commanded battleships in World War II. Even with the pessimistic decrement in bombing, Lee claimed a twenty-four-plane strike might still get as many as six hits on a light cruiser.

79. Ibid., 230–232.

80. See also Miller, 177–179. Moffett's successor, Admiral Ernest King, in addition to supporting the flying deck cruiser idea brought to fruition the acquisition, in 1933, of the Consolidated PBY-Catalina that proved invaluable in the first year of the war against the Japanese.

81. PHGB, December 5, 6, and 23, 1930. Zimm, 235–236. Zimm cites a BuC&R design dated June 15, 1931, No. 011066. The design bears a striking resemblance to the later *Kiev*-class flying-deck "cruisers" built by the Soviet Union during the Cold War as sea-control ships. See also Friedman, *U.S Cruisers*, 174–175.

82. Robert W. Love, Jr., *History of the U.S. Navy, Volume II*, 1942–1991 (Harrisburg, PA: Stackpole Books, 1992), 378–379.

83. GB 420-2 Serial No. 1523, "Building Program 1933."

84. PHGB, July 16 1931. Secretary of the Navy Adams was the most involved secretary in the hearings of the General Board during the interwar period.

85. PHGB, July 16, 1931. Admiral Rock pressed Bristol for an immediate decision, venturing that the ship might be built in less than thirty months if they acted quickly. The meeting held was not transcribed and probably had to do with placement of the AA guns and their directors.

86. GB 438-1 December 5, 1931 memorandum for the Secretary of the Navy. The Navy learned of this proposal due to its informal participation in the League Preparatory Commission meetings that summer. It was formally adopted by the League on September 29, 1931. The Board drafted a serial recommending the United States not ascribe to this proposal. See Chapter 4, for this discussion. President Hoover ignored the Board's advice and applied the League's moratorium to U.S. naval construction. GB 420-2 October 1, 1931, memorandum for the chief of naval operations from the General Board by T. C. Kinkaid, cites the "proposed truce" as the reason 1932 funds were not programmed for the flying deck cruiser.

87. Zimm, 233; PHGB, December 23, 1930. Zimm gives these characteristics as being designated A2, they clearly were an evolutionary development from the D-designs discussed during the hearings on December 23. The H designs were proposed in January and were considered too lightly armored by BuC&R and this was probably what the General Board discussed during its meeting on January 23, 1931.

88. Zimm, 242. GB 420-2 January 2, 1932 memorandum from the office of the chief of naval operations to the General Board. This memo referenced the bill cited, which allotted fifteen million dollars for one CF in 1933 (keel laid), and two each for 1935, 1936, and 1937. Serial 1568, April 13, 1932, the revised building program request for 1933, reflected only one CF.

89. GB 438, copy of May 1, 1933 "Memo of Admiral Hilary Jones for Navy League." See also Zimm, 242.

90. PHGB roll 1, index of membership; Zimm, 243.

91. See Friedman, *U.S. Cruisers*, 173.

92. PHGB, January 31 and February 15, 1934, "Flying Deck Cruiser;" Zimm, 244; and Friedman, 134, 174.

93. Duncan S. Ballantine, *U.S. Naval Logistics in the Second World War* (Princeton: Princeton University Press, 1947), 29. Ballantine characterizes the Navy as being "caught between the Unites States' avowed policy of disarmament of the one hand and an economy-minded Congress on the other." He lays most of the blame at the door of Congress.

Chapter 7

Note: The epigraph was taken from PHGB, December 3, 1923, "Characteristics of Floating Drydocks."

1. Dry docks had become the critical logistic support equipment for ships in the industrial age. They were essentially large floodable basins that could then be drained of their water to allow complete access to a ship's hull. They vastly improved the longevity and maintenance of the hulls of ships. Especially important was how marine biology affected ship's speed over time and the need to clean hulls after extended cruising of anything beyond six months. Speed, as at Coronel in World War I, could make all the difference between victory or defeat. Too, repair for below-waterline damage could only properly and quickly be done using a dry dock. Major repair to a ship's screw could often only be done in a dry dock.

2. RG80, Chief of Naval Operations Correspondence, 198-1, 1919–1927, Appendix F to WPL-9, December 1923 Mobile Base Project (hereafter RG80, 1923 MBP).

3. Bureau of Yards and Docks, *Building the Navy's Bases in World War II: History of the Bureau of Yards and Docks and Civil Engineer Corps, 1940–1946, Volume II* (Washington, D.C.: U.S. Government Printing Office, 1947), 228–231.

4. H.H. Frost, W.S. Pye, and H.E. Yarnell, "The Conduct of an Oversea Naval Campaign" (Washington, D.C.: Office of Naval Intelligence, October 1920), passim.

5. NARA, General Board (GB), Serial No. 1162, April 7,1923.

6. GB Serial No. 1088, 20 "Part IV—second section, Fortifications of Oahu, Guam and Philippines."

7. Bruce W. Menning, "Neither Mahan nor Moltke: Strategy in the Russo–Japanese War," in *The Russo–Japanese War in Global Perspective: World War Zero,* eds. John W. Steinberg, Bruce W. Menning, David Van Der Oye, David Wolff, and Shinji Yokote (Boston: Brill, 2005), 154–156. See also Constantine Pleshakov, *The Tsar's Last Armada* (New York: Basic Books, 2002), *xv–xviii*.

8. Japanese command of the sea during the Russo–Japanese War revolved around the ability of the Russians to dispute command of the sea from their only base in the theater of operations, Port Arthur, which was situated strategically on the Yellow Sea at the tip of the Liaodong peninsula. Without this base the Russians had no way to significantly influence events at sea. See also Menning, 129–156.

9. "Reconstitute" is a term from U.S. military doctrine that means to regain lost combat power. In the case of ships this means repairing damage, refueling, and accomplishing maintenance not possible at sea.

10. See Miller, Chapter 6 and passim for an overview of the impact of the Washington Conference on Orange planning. See also Evans and Peattie, 188–189. They identify the problem of retaining the Philippines as "nearly insoluble" prior to the fortification clause, which effectively made the problem even harder than it already was.

11. NARA RG38, War Plans Division, WPL-9, "Basic Orange Plan," February 1924, Chapter III, Basic Logistics Plan, 129 and http://www.chinfo.navy.mil/navpalib/ships/battleships/bbhistory.html, accessed July 5, 2006.

12. GB 420 serials 1162, 1271, 1305, 1338, 1376, and 1415, and so on, a recommendation for floating dry dock funding was missing from serial 1338 of December 1926. This is a reflection of the paucity of funds for naval construction prior to the Geneva Conference.

13. Ballantine, 16–17.

14. Bureau of Yards and Docks, *Building the Navy's Bases in World War II: History of the Bureau of Yards and Docks and Civil Engineer Corps, 1940–1946, Volume I* (Washington, D.C.: U.S. Goverment Printing Office, 1947), 209.

15. Frost, Pye and Yarnell, 17, 25, 44, 47. Marine railways are a useful way to load smaller ships onto flatbeds designed for underwater use which can then be used to bring these ships out of the water for hull maintenance in the open air.

16. Alan R. Millett, "Assault from the Sea," in *Military Innovation in the Interwar Period*, eds. Alan R. Millet and Williamson Murray (Cambridge, U.K.: Cambridge University Press, 1996), 71–73. Ellis had written "Advanced Base Force Operations in Micronesia" in July 1921 wherein he posited the requirement for seizing Japanese-held islands to provide intermediate Pacific bases for the Navy. The authorship of two studies so overwhelmingly concerned with advanced base operations so close in time, Ellis's following that of Frost et al., suggests that the one document, vague and nonspecific, was the template for its more specific successor.

17. See Miller, 83 and 114 and Baer, 100–102, 117–118. Baer emphasizes that with the advent of the fortification clause any hope to execute the pre-Washington Conference Orange Plan had vanished.

18. Miller, 114–115. According to Miller, Williams was nicknamed "The Oracle" during his stint on the General Board based on "his grasp of the future."

19. PHGB, December 3, 1923, "Characteristics of Floating Drydocks," and GB 420-2 Serial 1162. Coffey referred to the mobile base initiative as emerging "about a year ago" which would date the idea's genesis as late 1922/early 1923. The Navy was not consistent in its usage of the term "dry dock," sometimes listing it as two words, sometimes, as in this case, as one word "drydock."

20. PHGB, December 3,1923.

21. Ibid.

22. NARA RG80 Correspondence of the Chief of Naval Operations, War Plans S.C. 192-1, "Basic Readiness Plan, Appendix F—Mobile Base Project," December 20,1923, 1 (hereafter 1923 Mobile Base Project).

23. 1923 MBP, 4.

24. Ibid., 10–13. The General Board referred to these docks as "types" and OpNav referred to them as classes.

25. PHGB, January 3, 1924, "Characteristics of Floating Dry Docks," 15. See transcript cited later on this chapter.

26. NARA RG38, office of the chief of naval operations, December 1, 1927, "Synopsis of a Preliminary Technical Study—Basic Orange Plan, Advanced Base," 4.

27. PHGB, January 3, 1924, "Characteristics of Floating Dry Docks," 13.

28. PHGB, January 15,1925, 16.

29. YFD stands for Yard Floating Dock. The terminology for the mobile dry docks was to be ARD, auxiliary floating (repair) docks. They were not considered ships per se, although they had a commanding officer and crew assigned. Often the crew lived ashore or on a tender.

30. PHGB, January 15 1925, 13 and passim; PHGB, January 3, 1925, passim.

31. Ibid., 17 and passim; PHGB, January 3, 1925, passim.

32. PHGB, January 3, 1925, 15.

33. NARA RG80, GB 420-2, February 16, 1925 memorandum from the president to the General Board. This correspondence no longer listed the floating dry dock recommended in the earlier serial (1162) on the president's authorization request for the next fiscal year.

34. PHGB, March 25, 1925, "Characteristics of Floating Dry Docks."

35. A ship's draft is the depth of water the ship displaces below the waterline.

36. NARA RG38, office of the chief of naval operations, September 23, 1936, "Class A and Class B Floating Dry Docks—Brief History of the Evolution and Present Status of ARD-3." (hereafter OpNav September 23, 1936) This secret memo was part of a package submitted by War Plans to the Chief of Naval Operations to secure additional funding support to build the largest floating dry dock, now categorized as type B and included the final design description as "shipshape gate type." It included a brief history of the entire floating dry dock program.

37. PHGB, March 25 and 30, 1925, "Characteristics of Floating Dry Docks"; OpNav September 23, 1936.

38. PHGB, March 25,1925.

39. Ibid., 13–14. BuC&R's letter was read in its entirety into the hearing transcript.

40. PHGB, March 25 and 30 and April 2, 1925. GB 420-2 Serial No. 1271, April 3, 1925 and GB 420-2 Serial No. 1243, January 11, 1926.

41. OpNav, September 23,1936.

42. OpNav, September 23, 1936; PHGB, January 29, 1936, "Dry Dock ARD-1." The 1936 Board hearing included an extended discussion of how ARD-1 came to be built.

43. NARA RG38, correspondence of the chief of naval operations, 01 December 1927, "Synopsis of a Preliminary Technical Study—Basic Orange Plan, Advanced Base." The only annotation on the face of this study was handwritten by Admiral Schofield directing Captain Horne's attention to the "earliest dates to complete floating dry docks page 4."

44. PHGB, January 1 and 29, 1935, "Dry Dock ARD-1," September 13, 1935, "Floating Dry Dock ARD-3," and January 31, 1936, "Characteristics of Floating Dry Dock ARD-2"; OpNav September 23, 1936.

45. OpNav, September 23, 1936; NARA RG38, Correspondence of the chief of naval operations, February 19, 1936, "Master Priority List, Dry Dock Items, Relative Importance under the Orange Plan." The total number of docks envisioned in this list was 15, including 2 Class A, the Class B were to be increased in size to accommodate the carriers of the *Yorktown* class.

46. PHGB, September 13, 1935. These comments were on the last page of the Board's transcripts and were made by Cdr. Carroll of the War Plans Division.

47. NARA RG38, Box 42, office of the chief of naval operations secret memorandum, February 12, 1937.

48. NARA RG80, GB 420-2, Bureau of Yards and Docks, "Shipbuilding and Docking Facilities," May 1, 1940.

49. http://www.navsource.org/archives/09/28idx.htm, accessed August 15, 2006.

50. http://www.navsource.org/archives/09/2801.htm, accessed August 22, 2006; Ballantine, 161–162; Worral Reed Carter, *Beans Bullets, and Black Oil: The Story of Fleet Logistics Afloat in the Pacific during World War II* (Washington, D.C.: U.S. Government Printing Office, 1953), *vii*–ix, 54–55. Admiral Carter was essentially the Mobile Base's first commander for the Solomons Campaign based out of Espiritu Santo and later commanded Service Squadron Ten during the war—he was a battleship sailor. His memoir is a testament to the Navy's transformation of some officers from battleship warriors into logistics warriors. The preface to the book, by Admiral Raymond Spruance, is also a testament to profound change toward mobile logistics by the Navy's executive leadership.

51. http://freepages.military.rootsweb.com/~cacunithistories/Dewey_Drydock.html; http://www.ibiblio.org/hyperwar/USN/ships/ships-ar.html. Both links accessed August 24, 2006.

52. Ballantine, 162, 287.

53. Evans and Peattie, 290.

54. Vernon Clark, "Sea Power 21: Projecting Decisive Joint Capabilities," *United States Naval Institute Proceedings* (October 2002), 32. Clark, then Chief of Naval Operations, identified three "fundamental concepts [that] lie at the heart of the Navy's continued operational effectiveness: Sea Strike, Sea Shield, and Sea Basing." Note the similarity to Mahan's and Schofield's conception structurally with three elements; note, too, the prominence of positioning logistics at sea.

Chapter 8

Note: The epigraph was taken from Asada, "From Washington to London," 154; Stephen Roskill, *Naval Policy Between the Wars, vol. I* (New York: Walker and Co., 1968), 330.

1. John Terraine, *U-Boat Wars, 1916–1945* (New York: G. P. Putnam's Sons, 1989), 669.

2. Article 183 of the Versailles Treaty limited the *Reichsmarine* to a total of 15,000 personnel of which only 1,500 could be officers.

3. Corbett, p. 234. Corbett identifies such a fleet as one that is "—not merely in existence, but in active and vigorous life."

4. Terraine, 154.

5. Ibid., 68, 99. Japan provided destroyers for ASW escort duty for Allied shipping in the Mediterranean.

6. William S. Johnson, "Naval Diplomacy and the Failure of Balanced Security in the Far East: 1921–1935," *Naval War College Review*, 24 No. 6 (February 1972), 70.

7. Ibid., 70–71.

8. Terraine, 157–158.

9. Williamson Murray, "Strategic Bombing: The British, American, and German experiences," in W. Murray and A. R. Millet, eds., *Military Innovation in the Interwar Period* (Cambridge: Cambridge University Press, 1996), 101–106. "

10. Roskill, *Naval Policy Between the Wars, Vol. II*, 188–189, 238, 338. British attitudes which attributed Japanese inferiority to racial factors were still promulgated as orthodox and reasonable as late as 1935 to the Staff College.

11. Murray and Millett, Chapters 5 and 6. In chapter 5 Geoffrey Till criticizes British development of naval aviation. In Chapter 6 Holger Herwig included the British in his general criticism of all interwar navies as to their doctrine for antisubmarine and submarine warfare.

12. Holger Herwig, "The Submarine Problem," in Murray and Millett, eds. *Military Innovation in the Interwar Period,* 245. ASDIC stands for Anti-Submarine Detection Investigation Commission.

13. Terraine, 184.

14. See Hone et al., 188–195. See also Geoffrey Till, "Adopting the Aircraft Carrier," 208, and 191–226, passim. Till terms control of British Naval Aviation "dual control" due to air ministry control of Royal Navy aviators' careers and naval aviation resources and budgets.

15. Till, 199.

16. Roskill, 258.

17. Evans and Peattie, p.129. The authors trace this to the "myths" generated by Japan's victory at Tsushima. Asada, "From Washington to London," 150. Asada discusses Admiral Togo's "reverence" for Mahan.

18. Asada, 183–184.

19. Asada, "The Revolt against the Washington Treaty: The Imperial Japanese Navy and Naval Limitation, 1921–1927," *Naval War College Review, Summer 1993* (Newport, RI: NWC Press, 1993), 89; Asada, "From Washington to London," 153.

20. Ibid., 84.

21. Evans and Peattie, 515–516. Asada, 86. Asada, too, highlights the great paradox between Japan's strategic means and ends.

22. Johnson, 83. Johnson outlines how the Japanese Navy militants were never satisfied with the Naval budget and how Japan built while the United States and Britain remained idle. The United States, after Japan's conquest of Manchuria, passed the Vinson-Trammel Act in 1934 that authorized building to treaty limits in response to the Japanese hardliners.

23. *Kaga* was actually converted from a battleship hull.

24. For the Japanese these were carriers like *Ryujo*, for the USN there was *Ranger*.

25. Till, Geoffrey, "Adopting the Aircraft Carrier: The British, American, and Japanese Case Studies," from Murray and Millet, *Military Innovation in the Interwar Period*, Chapter 5, passim.

26. Goldman, 293. "A Treaty Between the Same Powers, in Realtions to the Use of Submarines and Noxious Gasses in Warfare" was signed at Washington at the same time as the Washington Naval Treaty. It addressed the proper use of submarines against merchant vessels, to ensure the safety of the crew before sinking the vessel and must search the vessel to determine its character.

27. Mark Parillo, *The Japanese Merchant Marine in World War II* (Annapolis, MD: Naval Institute Press, 1993), passim. Parillo's indictment is scathing.

28. Johnson, 77–78.

29. Evans and Peattie, 206–207.

30. Ibid., 205–207. Asada, 93–94. The typical IJN destroyer squadron was composed of five or more destroyers commanded by an admiral who commanded from a light cruiser flagship.

31. Asada, 93. Kato Kanji was quoted as saying, "We must devote ourselves more and more to this kind of drill, to which our navy has applied all its energies ever since the acceptance of the 10:10:6 ratio."

32. G. H. Bennett and R. Bennett, eds., *Hitler's Admirals* (Annapolis, MD: Naval Institute Press, 2004), 41–43. After World War I the German navy retained its title as the *Reichsmarine* (Imperial Fleet). After Hitler came to power it was redesignated the *Kriegsmarine* (War Navy). Testimony of Admirals Heye, Boehm, and Kranke. This source consists of translated essays written by German Admirals—including Doenitz—for British Naval Intelligence teams in 1945–1946. See also Erich Raeder, *Grand Admiral* (Annapolis: Naval Institute Press, 1969), 266.

33. Corum, *Roots of Blitzkrieg,* passim. Corum's work highlights how the German army conceptualized and practiced combined arms mechanized warfare in the absence of resources—like tanks and airplanes.

34. John Keegan, *The Price of Admirality* (New York: Penguin Books, 1988), 222–224. Terraine, Part II, passim.

35. Keegan, 223–224.

36. Terraine, 186; Karl Doenitz, *Memoirs: Ten Years and Twenty Days*, trans. R. H. Stevens (Annapolis, MD: Naval Institute Press, 1959), 37–39. Doenitz' criticisms of the Z-Plan were clearly the result of hindsight. Nowhere does he address Raeder's purpose of using the fleet as a balanced, combined arms, commerce-raiding force. He also uses the polemic "so-called" in referring to the Z-Plan, something neither Raeder nor Ruge do in their accounts. Bennett and Bennett, 62–63. Doenitz' comments in these essays concerning surface ships are completely at odds with his later memoir. Also, Doenitz is very critical concerning Goering's undermining of German naval aviation before the war.

37. Keegan, 224. Terraine also makes these points in Part II of his book.

38. Terraine, 194. Doenitz, 30–31.

39. Terraine, 174.

40. Friedrich Ruge, *Der Seekrieg: The German Navy's Story, 1939–1945* (Annapolis: United Naval Institute Press, 1957), 29.

41. Terraine, p. 204.

42. Ibid., 165. To further quote from this work: " . . . the German Admirals (Raeder himself conspicuously) seemed unable to drag themselves away from big ships and big guns."

43. Ruge, 34–35. Erich Raeder, *My Life*, trans. H. W. Drexel (Annapolis: Naval Institute Press, 1969), 272–273. Ruge implies that Raeder always had a commerce war in mind that would integrate surface, subsurface, and air forces to the purpose at hand. The Royal Navy was to be defeated only as a means to help strangle Britain's sea-based communications. Raeder's memoirs support this conclusion, citing German naval staff studies used in the design of the fleet.

44. Ruge, 34–38.

45. Ibid., 46–50.

46. Morison, Chapter 1 passim; Terraine, 157. Morison and Terraine are the principal claimants here for the case of America and for Germany/Great Britain, respectively.

47. Bennett and Bennett, 126. Kranke, Doenitz, Meyer, and a host of other German Admirals provide abundant evidence for this position.

Chapter 9

Note: The epigraph was taken from From Russell F. Weigley, *The American Way of War: A History of United States Military Strategy and Policy* (Bloomington, IN: Indiana University Press, 1973), 265.

1. From the definition of "ideology" found in *Webster's II, New Riverside University Dictionary* (Boston: Houghton Mifflin Company, 1988), 697.

2. Norman Friedman, *U.S. Destroyers,* Revised Edition (Annapolis, MD: Naval Institute Press, 2004), 76–77, 463–464; and Friedman, *Submarine Development and Design*, 76–77.

3. See Geoffrey Till, *Naval Transformation, Ground Forces, and the Expeditionary Impulse: The Sea-Basing Debate* (Carlisle, PA: U.S. Army War College, Government Printing Office, 2006) for a recent discussion of the Navy's sea-basing concept. The original concept was outlined in Vernon Clark, "Sea Power 21: Projecting Decisive Joint Capabilities," *United States Naval Institute Proceedings* (October 2002), 32. Clark, then Chief of Naval Operations, identified three "fundamental concepts [that] lie at the heart of the Navy's continued operational effectiveness: Sea Strike, Sea Shield, and Sea Basing."

4. Knox, *The Eclipse of American Sea Power*, 117, 131–132. Proceedings and Hearings of the General Board, March 3, 1922, 26.

5. GB Serial No. 1083, July 15, 1921. Adm. W. L. Rodgers, author of this serial, was perhaps the most articulate proponent of this view. Later Admirals like Hilary Jones and even William Pratt, after the disappointments in building the treaty navy, continued to echo them.

6. Murray and Millett, 313–314. The term "lessons learned" is charged with danger for historians. Nevertheless, it resonates with military audiences. I have chosen to replace it, where appropriate, with the more useful word "insight."

7. Mark Peattie, *Sunburst: The Rise of Japanese Naval Air Power, 1909–1941* (Annapolis: Naval Institute Press, 2001), 17. *Hosho* was laid down as a carrier from the keel up. *Langley*, as already discussed, was a converted collier.

8. Mahan, "Reflections, Historic and Other, Suggested by the Battle of the Japan Sea," USNI Proceedings, Vol. 23, No. 2 (June 1906), 469–471. Mahan writes a few pages later that

" . . . in every class of naval vessel there should first of all, and first and last, throughout her design, be the recognition of her purpose in war."

9. As noted earlier in this study, this tentative conclusion has also been suggested by Emily Goldman, among others, in *Sunken Treaties*.

10. The Joint Staff refers to the staff of the Chairman of the Joint Chiefs of Staff of the United States. The Joint Staff establishes the context within which the U.S. Navy conducts predeployment exercises; see http://www.news.navy.mil/search/display.asp?story_id= 24755, accessed April 10, 2007.

11. Asada, "From Washington to London," 183–184. Asada also uses the term "command faction" to refer to the heirs of Admiral Kato Kanji.

12. Hone et al., 173–175; Till, in Williamson and Millett, *Military Innovation during the Interwar Period*, Chapter 5, passim. Hone et al. identify USN organizational advantages in comparison to the Royal Navy. Till makes the explicit arguments about weaknesses in the "divided command" arrangements adopted by the British for their naval aviation.

13. Willmott, 3. "The Imperial Japanese Navy was checked and fought to a standstill, to a point of mutual exhaustion, by the prewar U.S. Navy. . . . "

14. Clark, 32.

Appendix 1

1. Access to these tables can be found in the *Encyclopedia of Arms Control and Disarmament*, Volume III or online at http://www.ibiblio.org/pha/pre-war/1922/nav_lim.html.

Appendix 3

1. Serial No.1121, April 26, 1922.

Appendix 4

1. From BuAer, December 29, 1930. Attached to RG80 NARA, PHGB, "Military Characteristics of Cruisers with Landing-on Decks," December 23, 1930.

Bibliography

Primary Sources

Contemporary Professional Journals

United States Naval Institute Proceedings
Naval War College Review
Military Review
Brassey's Naval Review
Army Navy Journal

Archival Sources

Hoover Library, West Branch, Iowa. Commerce Paper Series: Box 8 Naval Investigation of 1920; Box 156 Limitations of Armaments Conference. Cabinet Office Series: Boxes 35–36, Policy messages, CNO reports, condition of the Navy, General Board, etc.; Boxes 38/39, Naval Conferences, Reduction, Intelligence, and Economy.

National Archives and Record Administration (NARA), Washington D.C., and College Park, MD. Proceedings and Hearings of the General Board of the U.S. Navy, 1900–1950 (microfilm). [This microfilm collection is also located at Fort Leavenworth, Combined Arms Research Library.] Record Group 80, General Board Studies, 420, 420-2, 449, 438, 438–1, 438–2; Record Group 80, Chief of Naval Operations Correspondence (microfilm), 198–1; Record Group 38, Formerly Classified Correspondence of the Office of Naval Intelligence; Record Group 38, Office of Chief of Naval Operations, War Plans Division; Record Group 80, Chief of Naval Operations Correspondence, RG80 (microfilm) Appendix F, WPL-9 Mobile Base Project, 20 December 1923; Record Group 38, February 1924 Basic Orange Plan WPL-9 (hereafter 1924 Orange).

Naval Historical Center (NHC), Navy Yard, Washington, D.C., Naval Warfare Division collections and Navy Department Library. Unpublished list of the members of the General Board from January 1919 to August 1938, provided by Ms. Cathy Lloyd,

Senior Archivist, Naval Historical Center. This working aid was generated from the proceedings (minutes) of the General Board; Papers of Thomas C. Kinkaid.

Other Primary Sources

Anderson, Walter S. Captain. "Limitation of Naval Armament," First Honorable Mention for U.S. Naval Institute Essay Contest, 1926. *Proceedings*, No 277, March 1926.
Bennett, G. H. and R. Bennett, eds., *Hitler's Admirals*. Annapolis, MD: Naval Institute Press, 2004. [Note, these are essays by ten different German Admirals written shortly after the end of World War II with editorial comments by the Bennetts.]
Buell, Raymond L. *The Washington Conference.* New York: D. Appleton & Co., 1922.
Bywater, Hector C. "The Battleship and its Uses," Prize Winner for U.S. Naval Institute Essay Contest, 1926. *U.S. Naval Institute Proceedings*, No 277, March 1926, 407.
Davis, George T. *A Navy Second to None: The Development of American Naval Policy.* New York: Harcourt Brace & Co., 1940.
Doenitz, Karl. *Memoirs: Ten Years and Twenty Days* translated by R.H. Stevens in collaboration with David Woodward; with an introduction and afterword by Jurgen Rohwer. Annapolis, Md.: Naval Institute Press, 1990.
Fiske, Bradley A., "The Relative Importance of the Philippines and Guam," *United States Naval Institute Proceedings*, Special Supplement to No. 225, November 1921, 1676-1 through 1676-3.
Grow, H. B. "Bombing Tests on the 'Virginia' and 'New Jersey,'" *U.S. Naval Institute Proceedings* Vol. 49, No. 12 (December 1923), 1, from a Navy Reprint Pamphlet.
Knox, Captain Dudley W., *The Eclipse of American Sea Power*. New York: American Army and Navy Journal, Inc., 1922.
Mahan, A. T. *The Influence of Sea Power Upon History*. New York: Hill and Wang, 1957, originally published in 1890.
———. *Naval Strategy*. Newport, RI: Department of the Navy, GPO, originally published 1909, reprint 1991 as U.S. Marine Corps pub *FMFRP 12–32*.
League of Nations, *Armaments Year—Books: General and Statistical Information*. Geneva, 1924–1936.
Pratt, William V. "Some Considerations Affecting Naval Polity," *United States Naval Institute Proceedings,* Volume No. 237, November 1922.
———. "Some Aspects of our Air Policy," Second Honorable Mention for U.S. Naval Institute Essay Contest, 1926. *United States Naval Institute Proceedings*, No. 277, March 1926.
Raeder, Erich. *Grand Admiral*, Annapolis: Naval Institute Press, 1969.
Sprout, Harold and Margaret, *Toward a New Order of Sea Power: American naval Policy and the World Scene, 1918–1922*. Princeton, NJ: Princeton UP, 1940.
Talbot, Melvin F. "The Battleship: Her Evolution and Her Present Place in the Scheme of Naval War," Prize Essay 1938, *United States Naval Institute Proceedings*, Vol. No. 64, No. 5, May 1938.

Secondary Sources

Books

Asada, Sadao. *From Mahan to Pearl Harbor: The Imperial Japanese Navy and the United States*. Annapolis, MD: Naval Institute Press, 2006.

Baer, George W., *100 Years of Sea Power: The U.S. Navy, 1890–1990*. Stanford, CA: Stanford University Press, 1994.

Buckley, Thomas H. *The United States and the Washington Conference, 1921–1922*. Knoxville, TN: University of Tennessee Press,1970.

Burns, Richard Dean, ed., *Encyclopedia of Arms Control and Disarmament, Volumes I–IIII*. New York: Charles Scribner's Sons, 1993.

Corbett, Julian S. *Some Principles of Maritime Strategy*. Reprint. Annapolis, MD: Naval Institute Press, 1988.

Corum, James S. *The Roots of Blitzkrieg: Hans von Seeckt and German Military Reform*. Lawrence, KS: University Press of Kansas, 1992.

Crowley, James B. *Japan's Quest for Autonomy: National Security and Foreign Policy, 1930–193*. Princeton, NJ: Princeton University Press, 1966.

Evans, David C. and Mark R. Peattie. *Kaigun*. Annapolis: Naval Institute Press, 1997.

Friedman, Norman. *U.S. Battleships: An Illustrated Design History*. Annapolis, MD: Naval Institute Press, 1985.

———. *U.S. Destroyers*, Revised Edition. Annapolis, MD: Naval Institute Press, 2004.

———. *Submarine Development and Design*. London: Conway Maritime, 1984.

———. *U.S. Submarines: Design and Development*. Annapolis, MD: Naval Institute Press, 1984.

———. *U.S. Cruisers: An Illustrated Design History*. Annapolis, MD: Naval Institute Press, 1984.

———. *U.S. Aircraft Carriers: An Illustrated History*. Annapolis, MD: Naval Institute Press, 1983.

Friedman, Norman, Mark Mandeles and Thomas C. Hone. *America and British Aircraft Carrier Development, 1919–1941*. Annapolis: Naval Institute Press, 1999.

Goldstein, Erik and John Maurer, eds. *The Washington Conference, 1921–22: Naval Rivalry, East Asian Stability and the Road to Pearl Harbor*. U.K.: Frank Cass, 1994.

Goldman, Emily O. *Sunken Treaties: Naval Arms Control Between the Wars*. University Park, PA: Pennsylvania State University Press, 1994.

Hammond, James. *The Treaty Navy: The Story of the U.S. Naval Service Between the World Wars* (Victoria, Canada: Trafford, 2001).

Hone, Thomas C. and Trent Hone. *Battle Line: The United States Navy, 1919–1939*. Annapolis, MD: Naval Institute Press, 2006.

Howard, Michael. *War and the Liberal Conscience*. New Brunswick, NJ: Rutgers University Press, 1994.

Huntington, Samuel. *The Soldier and the State*. Cambridge, MA: Harvard University Press, 1957.

Hyde, Harlow. *Scraps of Paper: The Disarmament Treaties Between the World Wars.* Lincoln, NE: Media Pub., 1988.
Karsten, Peter. *The Naval Aristocracy: The Golden Age of Annapolis and the Emergence of Modern American Navalism*, New York: The Free Press, 1972.
Kaufman, Robert G. *Arms Control During the Pre-Nuclear Era: The United States and Naval Limitation Between the Two World Wars.* New York: Columbia University Press, 1990.
Keegan, John. *The Price of Admiralty.* New York: Penguin Books, 1988.
Knisley, Ronald William. *The General Board of the United States Navy—Its Influence on Naval Policy and National Policy.* Thesis (M.A.). University of Delaware, 1967.
Knox, Macgregor and Williamson Murray, eds. *The Dynamics of Military Revolution, 1300–2050.* Cambridge U.K.: Cambridge UP, 2001.
Lepore, Herbert P. *The Politics and Failure of Naval Disarmament, 1919–1939*: *The Phantom Peace.* Queenstown, Ontario: The Edward Mellen Press, 2003.
Love, Robert W. *History of the U.S. Navy, Vols. I and II, 1775–1991.* Harrisburg, PA: Stackpole Books, 1992.
Mandeles, Mark D. *The Future of War: Organizations as Weapons.* Washington, D.C.: Potomac Books, 2005.
McBride, William M. *Technological Change and the United States Navy, 1865–1945.* Baltimore, MD: The Johns Hopkins University Press, 2000.
McKercher, B.J.C. Ed., *Arms Limitation and Disarmament: Restraints on War, 189–1939.* Westport, CT: Praeger, 1992.
Melhorn, Charles M. *Two-Block Fox: The Rise of the Aircraft Carrier, 1911–1931.* Annapolis: Naval Institute Press, 1974.
Miller, Edward S. *War Plan Orange.* Annapolis, MD: Naval Institute Press, 1991.
Morison, Samuel Eliot. *The Two-Ocean War: A Short History of the United States Navy in the Second World War.* Boston: Little, Brown & Co., 1963.
Moy, Timothy. *War Machines: Transforming Technologies in the U.S. Military, 1920–1940.* College Station, TX: Texas A&M University Press, 2001.
Murray, Williamson and Allan R. Millett, eds. *Military Innovation in the Interwar Period.* Cambridge: Cambridge University Press, 1996.
O'Connell, Robert L. *Sacred Vessels: The Cult of the Battleship and the Rise of the U.S. Navy.* Boulder, CO: Westview Press, 1991.
Parillo, Mark P. *The Japanese Merchant Marine in World War II.* Annapolis, MD: Naval Institute Press, 1993.
Poolman, Kenneth. *The Winning Edge: Naval Technology in Action, 1939–1945.* Annapolis, MD: Naval Institute Press, 1997.
Peattie, Mark R. *Sunburst: The Rise of Japanese Naval Air Power, 1909–1941.* Annapolis, MD: Naval Institute Press, 2001.
Pelz, Stephen E. *Race to Pearl Harbor: The Failure of the Second London Naval Conference and the Onset of World War II.* Cambridge, MA: Harvard University Press, 1974.
Posen, Barry R. *The Sources of Military Doctrine: France, Britain, and Germany Between the World Wars.* Ithaca: Cornell University Press, 1984.

Price, Scott T. *A Study of the General Board of the U.S. Navy, 1929–1933*. Thesis (M.A.). University of Nebraska at Omaha, 1989.

Reynolds, Clark G. *Command of the Sea: The History and Strategy of Maritime Empires*. New York: William Morrow and Co., 1974.

———. *The Fast Carriers: The Forging of an Air Navy*. New York: McGraw-Hill, 1968.

Rosen, Stephen Peter. *Winning the Next War: Innovation and the Modern Military*. Ithaca: Cornell University Press, 1991.

Roskill, W. Stephen. *Naval Policy Between the Wars: I The Period of Anglo-American Antagonism, 1919–1929*. New York: Walker and Co., 1968.

———. *Policy Between the Wars: II The Period of Reluctant Rearmament, 1930–1939*. Annapolis, MD: Naval Institute Press, 1976.

Ruge, Friedrich. *Der Seekrieg: The German Navy's Story, 1939–1945*. Annapolis: United State Naval Institute Press,1957.

Spector, Ronald. *Professors at War: The Naval War College and the Development of the Naval Profession*. Newport, RI: Naval War College Press, 1977.

———. *Eagle Against the Sun* (New York: Vintage Books, 1985).

———. *At War At Sea: Sailors and Naval Combat in the Twentieth Century*. New York: Viking, 2001.

Schroeder, Paul W. *The Transformation of European Politics, 1763–1848*. Oxford, U.K.: Clarendon Press, 1994.

Sumida, Jon Tetsuro. *Inventing Grand Strategy and Teaching Command: The Classic Works of Alfred Thayer Mahan Reconsidered*. Baltimore, MD: Johns Hopkins University Press, 1997.

Terraine, John. *The U-Boat Wars, 1916–1945*. New York: G. P. Putnam's Sons, 1989.

Vlahos, Michael. *The Blue Sword: The Naval War College and the American Mission, 1919–1941*. Newport R.I., Naval War College Press, 1980.

Weigley, Russell F. *The American Way of War: A History of United States Military Strategy and Policy*. Bloomington, IN: Indiana University Press, 1973.

Willmott, H. P. *The Battle of Leyte Gulf: The Last Fleet Action*. Bloomington: Indian University Press, 2005.

Winton, Harold R. and David R. Mets, eds. *The Challenge of Change: Military Institutions and New Realities, 1918– 1941*. Lincoln: University of Nebraska Press, 2000.

Note: Copies of the treaties themselves are available as appendices in several of the secondary sources, especially the *Encyclopedia of Arms Control and Disarmament* and Goldman. They are also available online.

Government Documents and Studies

Headquarters, Department of the Army. *Operations*, FM 3-0. Washington, D.C.: U.S. Government Printing Office, June 2001.

———. FM 5-0 *Planning*. Washington D.C.: U.S. Government Printing Office, July 2002.

———. *Tentative Field Service Regulations—Operations*, FM 100-5. Washington D.C.: U.S. Government Printing Office, 1939.

———. *Field Service Regulations, United States Army*. Washington D.C.: U.S. Government Printing Office, 1923.

Kem, Jack D., "Strategic Concepts," from *C200 Strategic Studies Readings and Advance Sheets*. Fort Leavenworth, KS: U.S Government Printing Office, July 2005.

Lykke, Arthur F., Jr. "Toward and Understanding of Military Strategy," in *Military Strategy: Theory and Application*. Carlisle Barracks, PA: U.S. Government Printing Office, 1989.

Mahan, A. T. "The Battle of the Sea of Japan," USNI Proceedings, Vol. 23, No. 2, 471, 1906.

Mandeles, Mark D. The Jean de Bloch Group. *Developing New Tactics, Operational Concepts, and Organizations: The Interwar Period and Transformation to a Twenty-First Century Military*. Fairfax, VA: May 2003.

Naval Historical Center. *Dictionary of American Naval Fighting Ships, Volumes I–VIII*. James L. Mooney, Ed. Washington, D.C.: U.S. Government Printing Office, 1991. (Hereafter DANFS).

———. Web site http://www.history.navy.mil/ (Last accessed June10, 2008).

O'Neil, William D. *Transformation and the Officer Corps: Analysis in the Historical Context of the U.S. and Japan Between the World Wars* (draft copy). Alexandria, VA: Center for Naval Analysis, 2005.

———. *Interwar U.S. and Japanese National Product and Defense Expenditure*. Alexandria, VA: Center for Naval Analysis, 2003.

———. *Military Transformation as a Competitive Systemic Process: the Case of Japan and the United States Between World Wars*. Alexandria, VA: Center for Naval Analysis, 2003.

Secretary of the Navy Sean O'Keefe, CNO Admiral Frank B. Kelso II, Commandant USMC Gen C.E. Mundy, Jr. . . . *From the Sea: Preparing the Naval Service for the 21st Century*. Washington, D.C.: Department of the Navy, 1992.

Till, Geoffrey. *Naval Transformation, Ground Forces, and the Expeditionary Impulse: The Sea-Basing Debate*. Carlisle, PA: U.S. Army War College, Government Printing Office, 2006.

U.S. Department of Defense. Office of the Joint Chiefs of Staff. Joint Pub 3-0, *Doctrine for Joint Operations*. Washington, D.C.: U.S. Government Printing Office, 1995.

U.S. Senate Document No. 77. *Conference on The Limitation of Armament, Senate Document*. Washington, D.C.: U.S. Government Printing Office, 1921.

———. No. 126. *Conference on Limitation of Armament*. Washington, D.C.: Government Printing Office, 1922.

U.S. Government. *Papers Relating to the Foreign Relations of the United States: 1922*, Vol. 1, pp. 247–266. Treaty Series No. 671. http://www.ibiblio.org/pha/pre-war/1922/nav_lim.html (Last Accessed 10 June 2008).

Other Secondary Sources

Asada, Sadao, "From Washington to London: The Imperial Japanese Navy and the Politics of Naval Limitation, 1921–1930," in Erik Goldstein and John Maurer, Eds. *The Washington Conference, 1921–1922: Naval Rivalry, East Asian Stability and the Road to Pearl Harbor. U.K.: Frank Cass, 1994.

———."The Revolt against the Washington Treaty: The Imperial Japanese Navy and Naval Limitation, 1921–1927," *Naval War College Review*, Summer 1993. Newport RI: NWC Press, 1993: 82–96.

———. "Japanese Admirals and the Politics of Naval Limitation: Kato Tomosaburo vs Kato Kanji," In Gerald Jordan, ed., *Naval Warfare in the Twentieth Century, 1900–1945: Essays in Honor of Arthur Marder.* New York: Crane Russak, 1977.

———. "The Japanese Navy and the United States," in Dorothy Borg and Sumpei Okamoto, ed., *Pearl Harbor as History: Japanese-American Relations, 1931–1941.* New York: Columbia University Press, 1973.

———. "Japan's special Interest and the Washington Conference, 1921–1922." *American Historical Review* 67 (October 1961): 62–70.

Braisted, William R. "The Evolution of the United States Navy's Strategic Assessments in the Pacific, 1919–1931," in Erik Goldstein and John Maurer, eds. *The Washington Conference, 1921–1922: Naval Rivalry, East Asian Stability and the Road to Pearl Harbor,* U.K.: Frank Cass, 1994.

———."On the General Board of the Navy, Admiral Hilary P. Jones, and Naval Arms Limitation, 1921–1931," in *The Dwight D. Eisenhower Lectures in War and Peace, No. 5.* Manhattan, KS: Kansas State University, 1993.

———. *The United States Navy in the Pacific, 1909–1922.* Austin, TX: University of Texas Press, 1971.

Buckley, Thomas H. "The Washington Naval Limitation System: 1921–1939," in Richard Dean Burns ed., *Encyclopedia of Arms Control and Disarmament, Volume II.* New York: Charles Scribner's Sons, 1993.

———. "The Icarus Factor: The American Pursuit of Myth in Naval Arms Control, 1221–36," in Erik Goldstein and John Maurer, eds. *The Washington Conference, 1921–1922: Naval Rivalry, East Asian Stability and the Road to Pearl Harbor.* U.K.: Frank Cass, 1994: 124–146.

Clark, Vernon. "Sea Power 21: Projecting Decisive Joint Capabilities," *United States Naval Institute Proceedings* (October 2002), 32–41

Crowl, Philip A. "Alfred Thayer Mahan: The Naval Historian," in *Makers of Modern Strategy,* ed. Peter Paret. Princeton, NJ: Princeton University Press, 1986.

Heinrichs, Waldo H., Jr. "The Role of the United States Navy," in Dorothy Borg and Sumpei Okamoto, eds., *Pearl Harbor as History: Japanese-American Relations, 1931–1941.* New York: Columbia University Press, 1973.

Hone, Trent, "The Evolution of Fleet Tactical Doctrine in the U.S. Navy, 1922–1941," *Journal of Military History,* October 2003, 1107–1149.

Hone, Thomas and Mark Mandeles, "Managerial Style in the Interwar Navy: A Reappraisal," *Naval War College Review,* 32 (Sep–Oct, 1980), 88–101.

Johnson, William S. "Naval Diplomacy and the Failure of Balanced Security in the Far East: 1921–1935," *Naval War College Review,* 24 No. 6 (Feb 1972): 67–88.

Metz, Steven. "Strategic Asymmetry," *Military Review*, 81 (July–August 2001): 23–31.
Menning, Bruce W. "Neither Mahan nor Moltke: Strategy in the Russo-Japanese War," in *The Russo-Japanese War in Global Perspective: World War Zero*, John W. Steinberg, Bruce W. Menning, David Van Der Oye, David Wolff, and Shinji Yokote, eds. Boston: Brill, 2005: 129–156.
Milton, Keith. "Arctic Disaster," in *Military Heritage* (June 2001): 72–80.
Nye, Joseph S., Jr. and William Owen in *Foreign Affairs*, 75, No. 2 (March/April 1996): 20–36.
Price, Scott T. "A Study of the General Board of the U.S. Navy, 1929–1933." Unpublished master's thesis (M.A.), University of Nebraska at Omaha, 1989.
Rosen, Philip T. "The Treaty Navy, 1919–1937." In *In Peace and War: Interpretations of American Naval History, 1775–1984,* 2nd edition., ed. by Kenneth J. Hagan, 221–236. Westport, CT: Greenwood Press, 1984.
Semsch, Philip L. "Elihu Root and the General Staff," *Military Affairs* 27, no. 1 (Spring 1963), 16–27.
Spector, Ronald H. "The Military Effectiveness of the US Armed Forces, 1919–39," in *Military Effectiveness, Volume II The Interwar Period*, Eds. Allan R. Millett and Williamson Murray. Boston: Unwin Hyman, 1988, 70–97.
Sumida, John T. "Pitfalls and Prospects: the Misuses and Uses of Military History and Classical Military Theory in the 'Transformation' Era," unpublished draft, January 31, 2005.
———. "Information Dominance and Maneuver Warfare at Sea in the Early Twentieth Century: The Dynamics of Tactical Revolution and Reaction in the Dreadnought Era," Unpublished post-conference draft, October 25, 2004.
Trimble, W.F., "Admiral Hilary Jones and the 1927 Geneva Naval Conference," *Military Affairs* 48:3 (February 1979): 1–4.
Zimm, Alan D. "The U.S.N.'s Flight Deck Cruiser," *Warship International* No. 3, 1979, 216–245.

Index

Adams, C. F., 56, 98, 114
Advanced Base Sectional Dock (ABSD-1), 127, 142
"air depot ships." *See* tenders
aircraft carriers: conversion of battle cruisers to, 67; effect of Treaty on introduction of, 98–101; effects of budgetary constraints, 47–48, 49; *Ranger,* 48; and the reconstruction clause, 69–70; scaled-back size of, in priority program, 47–48; ship conversion allowance under Treaty, 42–43
Ajax, 92, 93–94
Akagi (Japan), 152
Akron, 122
American Historical Association, 33
American Society of International Law, 32–33
Anglo-French naval pact, 53
Anglo-Japanese Naval Alliance, 2, 146, 147, 159
Antietam, 118
Arizona class, 79
Arkansas, 73
Army Air Corps, 68
Army-Navy Joint Board, 17, 40
Army (U.S): Air Corps, 18, 68, 75, 175; importance compared to U.S. Navy, 177; proposal to use facilities for AF, 90, 95–98; Root Reforms of the, 8, 24
Artisan AFDB-1, 140
Asiatic Fleet (U.S.) (AF), 91; immediate effect of fortification clause on, 92; proposal to use Army facilities for, 90, 95–98; use of tenders in the, 92–98
Auxiliary (floating) Repair Docks (ARDs), 139–42
aviation. *See* naval aviation

Balfour, Arthur, 144
Barnett, Correlli, 147
battleship modernization, 86; aircraft spotting and range finding, 77–78; "Big Five," 75, 84; conversion of oil-burning to electric drives, 74; effect of Mitchell's bombing tests on, 68; focus on changes to existing designs, 64; gun armament, 65, 66; as highest priority in treaty fleet plan, 66; and the propulsion systems, 64–65, 66, 70–75; and the reconstruction clause, 69–70; target tests, 74–75; turret modernization to remedy short ranges, 78–81; turret modification funding problems, 81–82
battleships, 57; armor policy, 77; effect of Washington Naval Treaty on, 68–70; effect of World War I on importance of, 67; importance to Navy, 48; Strategic Problem IV, 84–85; submarines retained to sink, 68; viewed as final arbiter of naval power, 63–64
Beuret, J. D., 94
Beyerchen, Alan, 5
Bismarck (German), 157
bombing accuracy testing, 17, 68, 113, 122
Bristol, Mark: on design of flying deck cruisers, 110, 113, 114–15, 118–19; mentioned, 15, 109; questioning of naval air and fleet structure, 102; in relation to the Asiatic Fleet, 95–98
British and American Aircraft Carrier Development, 1919–1941 (Friedman, Hone, and Mandeles), 4–5
budgetary constraints and considerations, 25, 49, 163, 169–71; effect of World War I, 67; effect on aircraft carrier introduction, 95, 100–101; effect on battleship reconstruction, 25, 164; effect on U.S. Army, 90; effects on aircraft carriers, 47–48; and General Board

budgetary constraints and considerations (*continued*) recommendations for priorities regarding, 40; Hoover administration fiscal concerns, 98; Moffett on naval aviation neglect due to, 104; in relation to floating dry docks, 135–42

Building and Maintenance Policy of 1922, 69

Bureau of Aeronautics (U.S. Navy) (BuAer). *See also* Moffett, W. A.: and the Bureaus' joint recommendation for battleship modernization, 70–72; concerns about *Ajax*, 93–94; on naval aviation, 93, 96; questioning of naval air and fleet structure, 102; reactions to Whiting's report on aircraft spotting, 78; recommendation for battleship gun turrets modification, 72; in relation to flying deck cruisers, 110, 113–14, 118, 123; and U.S. naval aviation compared to Royal Navy, 149

Bureau of Construction and Repair (BuC&R): and the Bureaus' joint recommendation for battleship modernization, 70–72; in relation to floating dry docks and mobile bases, 136–39; in relation to flying deck cruisers, 109–10, 116, 118–19, 122

Bureau of Engineering (BuEng): and the Bureaus' joint recommendation for battleship modernization, 70–72; in relation to flying deck cruisers, 113, 114, 115

Bureau of Ordnance (BuOrd): on battleship modernization, 78, 79; and the Bureaus' joint recommendation for battleship modernization, 70–72; in relation to flying deck cruisers, 110, 113, 114, 115, 117; on U.S. Naval Policy, 36–37, 38

Bureau of Yards and Docks (BuY&D), 135–39, 141

Chief of Naval Operations office (CNO or OpNav), 17, 22. *See also* Orange War Plan; War Plans Division; *individual chief officer names*; in context of fortification challenges, 31; creation and focus of, 12, 13; on floating dry docks, 130–31, 133, 135, 139–41, 142; on flying deck cruiser design, 110; and U.S. naval aviation compared to Royal Navy, 149; utilization of the General Board's ideas, 12, 16

China "Open Door" trade policy, 29, 30

Clark, Vernon: and aircraft bombing testing, 122; on flying deck cruisers, 110, 115–16

Coffey, 132–33

Colorado, 84

Commander-in-Chief of the U.S. Asiatic Fleet (CINCAF), 93, 95–97, 100, 102

Commander-in-Chief of the U.S. Fleet (CINCUS), 35, 38, 100, 108

Coolidge, Calvin, 45, 75, 100–101

Coolidge administration, 44; cruiser limits in relation to Japan and Great Britain, 51–52; proposal of second "cruiser bill," 52

Coontz, R. E., 80–81

Corbett, Julian, 146

cruiser bills, 52, 53, 96

cruisers, 28, 50–56. *See also* flying deck cruisers; relative to the Reed-Matsudaira compromise, 55; spreading risk with flying deck, 55

Cunningham, A. C., 130

Curtiss, 98

Daniels, Joseph, 67

Denby, Edwin, 44, 81, 132

Dewey, 130, 136, 141, 142

Dewey, George, 10–11

Doenitz, Karl, 154–56

dry docks, described, 127

The Eclipse of American Sea Power (Knox), 32

Ellis, Earl H., 131, 134, 166

Far East, 27, 29, 90, 125, 134, 164; importance of fortification bases in, 31, 33, 73, 90, 105

Farragut class, 166

Fiske, Bradley A., 9–10, 30–31, 38, 174

Five Power Pact. *See* Washington Naval Conference; Washington Naval Treaty

Five Power Treaty. *See* Washington Naval Conference; Washington Naval Treaty

Fleet Base Force, 128, 129, 141, 166

Fleet Problem IX (1929), 40, 55, 103

floating dry docks, 43, 46, 47, 52, 56; budgetary constraints and considerations, 135–42; in context of the Office of Naval Intelligence study, 131; described, 127; development of the, 126–27, 128, 130

Florida, 73, 79

flying deck cruiser: linked to fortification clause and innovation, 123–24

flying deck cruisers: budgetary constraints, 118, 119, 121–22; compared to aircraft carriers, 103; in context of the 1930 London Naval Treaty, 102–3, 107; design changes after 1931, 122; design series for General Board consideration, 116–17; effect of League of

Nations moratorium, 119, 121; end of, 122–23; General Board hearings on design of, 109–10, 113–19; in the London Naval Treaty, 91; and Moffett's memorandum on, 104–5; redesignated as a "CF," 118
Ford Rangekeeper, 77–78
fortification clause (Article XIX), 4. *See also* General Board of the Navy; naval aviation; sea power problem; discussed in Williams' study, 56–57; Knapp on the, 32–33; Knox on impact of the, 32; and logistical workaround examples, 43; monetary effects of, 27; overview, 1; and potential western Pacific problems posed by, 31; as related to floating dry docks, 125, 126, 127, 131–32, 134–35, 137; in relation to battleship modernization, 65, 66, 84, 86–87; in relation to flying deck cruiser, 123–24; Schofield on the impact of the, 38
Four Power Pact, 2–3
Four Power Treaty, 147
"Four-Stackers," 52
France, 25, 59, 70, 147, 154
French, Burton, 100
Friedman, Norman, 4–5, 8
Frost, Holloway H., 130–31, 134, 140, 166

Gato class, 166
General Board of the Navy. *See also* fortification clause (Article XIX); treaty system; Washington Naval Treaty; *individual board member names*: 1927 prioritized program list, 46; aircraft tender considerations, 93–94; on battleship machinery and propulsion modernization, 72–75; collaboration with other Navy organizations, 12–13, 16; commitment to aircraft carriers versus flying deck cruisers, 108; construction programs curtailed after 1922, 44–45; effect of Chief of Naval Operations office creation, 12; floating dry dock considerations and hearings, 129, 132–33, 134, 135–42; and flying deck cruiser budgetary constraints, 118, 119, 121–22; and flying deck cruiser design, 103–4; flying deck cruiser design hearings, 109–10, 113–19; genesis of, 10–12, 24; hearing process, 16–18, 22; liberal interpretations of the Treaty, 40–41; linked to advising the Secretary of the Navy, 16–17, 40; members of the, 14, 15, 16, 21; Naval Policy "blueprints," 34–35, 35–36, 38, 40; opinion differences with civilian leadership, 42; opposition to the London Treaty, 55; overview and general role, 1, 8–10, 14, 15, 20–21; reactions to Whiting's report on aircraft spotting, 78; recommendations to Wilbur on naval limitation, 45–46; in relation to aircraft carrier introduction, 99–101; retention of dry docks in priority list, 46–47; Roosevelt's outlined areas for consideration by, 45; Serial 1239, 20, 45; Serial 1584, 57; Serial 1140 (U.S. Naval Policy) (*See U.S. Naval Policy*); Serial No. 1055, 41; serials or "studies," 17, 18–21, 41–42, 46, 47; summary of treaty system to 1932, 57; "treaty fleet" building program of 1929, 25; turret redesign debates, 78–81; and U.S. naval aviation compared to Royal Navy, 149; use of "second to none" axiom to support recommendations, 41–42; war plan function transferred, 14; Williams' outline for battleships, 85; "yellows," 17, 19
General Order 544, 11, 17
Geneva Conference of 1932, 56, 57
Geneva Naval Conference of 1927, 48, 50–52, 53
German Navy (*Kriegsmarine*): effect of Anglo-German London Naval Conference on, 155–56; effect of Treaty of Versailles on, 145–46; preparation for submarine war, 49, 155; use of U-boat, 155–56, 158–59; Z plan aimed at Great Britain, 154, 155, 156–58
Ghormley, Robert, 14
Gibson, Hugh S., 50
Goering, Hermann, 157
Graf Spee (German), 157
Great Britain. *See also* Geneva Naval Conference; London Naval Conference; London Naval Treaty of 1930: battleship and cruiser importance in Royal Navy, 150; in context of U.S. arms limitation, 45–46; events leading to Washington Naval Treaty, 145–46; factors effecting interwar naval innovation, 147–50; fiscal effect of Treaty of Versailles on Royal Navy, 146; linked to the reconstruction clause, 70; and naval aviation, 149–50, 177; protest of turret elevation, 81, 82; Royal Navy as compared to U.S. Navy, 149; submarine warfare, 148
Great Depression, 56, 98, 119, 122, 170
Gregory, L. E., 135–36, 138

Harding, Warren G., 45, 169–70
Harding administration, 25, 27, 42, 44–45, 146
Hart, Thomas, 15

Heinrichs, Waldo H., Jr., 8
Hermes (Great Britain), 97
Hitler, Adolf, 154, 155–56, 157, 158, 159
"holiday," treaty imposed: defined, 26; expiration of the, 82, 83, 165; extension considered, 20–21, 45, 76, 159; influence on ship construction, 39, 43, 55, 66, 69, 84, 86, 168
Hone, Thomas C., 4–5, 8
Hoover, Herbert, 10, 45, 53, 54–58, 139
Hoover administration, 10, 14, 17, 98, 122, 170
Hosho (Japan), 42, 172
Houghton, A. B., 82
House Resolution 8687, 47
House (U.S. colonel), 146
Hughes, Charles Evans, 25–26, 28, 33, 82, 101, 139; mentioned, 96
Hughes, Admiral C.F., 96
Huse, Harry, 73, 81

The Influence of Sea Power upon History (Mahan), 29
innovation, traditional and non-traditional described, 88–89
Italy, 25, 70

Japan, 32–33, 176–77. *See also* Asiatic Fleet (U.S.) (AF); Far East; fortification clause (Article XIX); Geneva Naval Conference; aircraft carrier/ship conversion allowance under Treaty, 42–43; Board's de-emphasis on Pacific strategy, 45–46; de-emphasis of Pacific area in the "Allocation Policy," 44–45; defiance of the Geneva Disarmament Conference of 1932, 57; effect of fortification clause on navy of, 153–54; effect of lack of auxiliary limits on, 28–29; events leading to Washington Naval Treaty, 145–46; impact of decisive battle focus, 150, 151–53; impact of perceived inferior position, 151–52; inclination toward budget reductions, 151; invasion of Manchuria, 57; naval aviation development, 152; Orange War Plan to locate fleet ships of, 117–18; and potential problems for U.S. posed by fortification clause, 31–33; and ratio system relative to U.S., 25, 26–27, 54–55; in relation to the 1930 London Conference, 38, 54–55, 151; submarine warfare, 152–53; tactical training of navy of, 153
Jason, 95–98
Joint Board, 63

Jones, Hilary P., 45, 48, 49–51, 54; as de facto "expert" on the Washington Naval Treaty, 72; on floating dry docks and mobile bases, 133, 135, 136–38, 142; mentioned, 96; opposition to flying deck cruisers, 91, 122; questioning of naval air and fleet structure, 102; support of turret modification funding, 81–82

Kago (Japan), 152
Kanji, Kato, 144, 151, 153
Kellogg-Briand Pact, 159
Kiev class (Soviet Union), 124
King, Ernest J., 14, 15, 123, 171
Kinkaid, Thomas C., 15
Knapp, H. S., 32–33
Knox, Dudley, 29, 32, 43, 167, 168; *The Eclipse of American Sea Power,* 32
Kriegsmarine. See German Navy
Kuhn, Thomas, 6

Langley, 42, 95, 97, 104, 110, 123, 172, 177
Laning, Harris, 108–9, 116–18, 122
League of Nations: disarmament conference in Geneva in 1932, 56; Mandate island groups, 27–28, 32; new construction moratorium, 119; Preparatory Commission for Disarmament, 48, 53
Leahy, William, 114–15, 141–42
Leary, H. F., 36–37
Lee, Willis, 117
Lejeune, John, 34
"levels" methodology (Beyerchen), 5
"levels of analysis" (Friedman et al.), 4–5
Lexington, 40, 47, 56, 91, 102, 106–7, 152; conversion to aircraft carrier, 99–101; and correction to design of flying deck cruisers, 110
Light Aircraft Carriers (CLVs), 117–18
Lodge, Henry Cabot, 28
London Naval Conference, 38, 52–56, 106–8, 113, 151, 164
London Naval Treaty of 1930, 10, 55–56, 58–59, 76, 148, 165; and naval aviation, 91, 102–3, 107–10, 164, 173
London Naval Treaty of 1935, 59, 148, 154, 156
Long, A. T., 15
Long, John D., 11
Lord, H. M., 98

Mahan, A. T.: early support for naval general staff, 10–11; inclusive approach to naval design, 53; *The Influence of Sea Power upon*

History (Mahan), 29; Japan adherence to principles of, 159; lectures at Naval War College, 30; *Naval Strategy: Compared and Contrasted with the Principles and Practice of Military Operations on Land,* 30; on operational effectiveness, 26; overseas bases as strategic positions, 30; sea power as framework for solution, 23, 25, 26, 29–30, 143, 150, 165, 174, 178–79; tactical maxim, 77
Mahan class, 166
Mandeles, Mark D., 4–5, 8
McNamee, Luke, 34
McVay, Charles B., 36, 79–80
Metcalf, M. K., 34
Michaelson, J., 138–39
Military Innovation in the Interwar Period (Murray), 4
Millett, Alan R., 131
Mitchell, Billy, 18, 68, 89
Mitscher, Marc, 113, 114–15
mobile base, origination of, 130–31. *See also* floating dry docks; mobile base project (MBP)
mobile base project (MBP), 125–26, 127, 132, 133–35, 139–40
Moffett, W. A., 72; on allocating budget for aircraft carriers, 100; on the Asiatic Fleet naval aviation situation, 95, 96–98; cruisers recommendation, 54, 55; death of, 122; on flying deck cruisers and aircraft carriers, 103–5, 106–8; on importance of CFs, 121; memorandum for Naval Policy, 36, 41; memorandum regarding aircraft building, 108; mentioned, 9–10, 19, 20; questioning of naval air and fleet structure, 102; support of naval aviation programs, 88, 90, 91; support of turret elevations, 82–83; on turret modernization to remedy short ranges, 78–81
Morison, Samuel Eliot, 159
Morrow Board of 1925, 89
Murray, Williamson, 4, 171

National Industrial Recovery Act (NIRA), 140
Naval Air Policy, 19, 20, 35, 36–37
naval aviation. *See also* aircraft carriers; Asiatic Fleet (U.S.) (AF): initiatives and challenges, 88–89; new dynamic of cruisers and aviation, 54
Naval Policy of 1922 (U.S.). *See U.S. Naval Policy*

Naval Strategy: Compared and Contrasted with the Principles and Practice of Military Operations on Land (Mahan), 30
Naval War College, 40, 90, 176; in context of fortification clause challenges, 31; creation of, 24; gaming as related to flying deck cruisers, 102, 109, 117, 122; Lieutenant Commander rank attending, 9; Mahan's lectures at, 30; president as member of General Board, 11, 12, 14, 34; utilization of the General Board's ideas, 12, 16; war gaming, 13, 15, 16, 31, 38
Navy General Order 94, 128
Navy Officer Corps, 9
Nevada class, 73–74, 82
New Jersey, 75
New Mexico class, 82
"New Orleans dock" (YFD-2), 130, 141, 142
New York, 73
Nicholson, (U.S. lieutenant commander), 110–11, 113, 114–15, 119
Nimitz, Chester, 162
Nine Power Pact, 2–3
1916 Navy Act, 67
1924 Orange War Plan document (WPL-9). *See* Orange War Plan
North Carolina class, 59, 64
North Dakota, 75

Office of Naval Intelligence (ONI): as member of General Board, 15; studies of the, 130–31, 134, 140, 166
OpNav. *See* Chief of Naval Operations office (CNO or OpNav)
Orange War. *See* Orange War Plan
Orange War Plan, 26–27, 31, 32, 38, 40, 57; on anxiety over Pacific problems, 76–77; battleship modernization to support, 64; dry docks, 47; and Light Aircraft Carriers (CLVs), 117–18; mobile base project (MBP) as part of, 34, 125–26, 127, 132, 133–35, 139–40; reliance on maximum naval construction, 102; replaced by Rainbow Plans, 177; role of flying deck cruisers in, 114; and the Strategic Problem IV, 84–85, 90; submarine significance in the, 49; "thrusting" strategy of the, 87, 90, 127, 128, 132, 133; and use of naval aviation, 89–90, 104–5
Ostfriesland (German), 68, 74

Pacific Fleet: advanced bases for, 126; importance of building carriers for, 100

Pacific problems. *See also* Asiatic Fleet (U.S.) (AF); Far East; Japan: overview of, 2, 3, 76–77

Pratt, 82; and flying deck cruisers, 108–10, 113, 115–16, 119, 121, 122; on aircraft carrier tonnage, 99, 101; on battleship modernization, 67; as battleship modernization technical advisor, 72; as de facto "expert" on the Washington Naval Treaty, 72; and floating dry dock construction, 139; on importance of naval aviation, 104; questioning of naval air and fleet structure, 102; support of aviation programs, 91

Pratt, William V., 9–10, 14, 16, 21, 47; as General Board member, 34; role at 1930 London Conference, 54–56, 58; as Washington Naval Treaty supporter, 33

Pye, William S.: in connection with Office of Naval Intelligence studies, 130–31, 134, 140, 166

radar development analysis, 5

radius of action, 43, 44, 46, 71, 73, 74; in relation to battleship modernization, 65, 66

Raeder, Erich, 154–58

Rainbow Plans, 175

Ranger, 48, 104

reconstruction clause, 65, 70–72, 74, 99, 163–64

Reed-Matsudaira compromise, 55

Reeves, J. M.: and floating dry dock design, 140; as opposition to flying deck cruisers, 122; questioning of naval air and fleet structure, 102

Renown (Great Britain), 70

Rodgers, William L., 18, 34–38, 73, 93, 106, 168; in relation to floating dry docks, 125, 135–37, 141–42

Roosevelt, Franklin D., 57–58, 122; and floating dry dock budget availability, 140

Roosevelt, Theodore, Jr., 20, 45; on retaining submarines to sink battleships, 68; as "war strategy board" member, 11

Ruge, Friedrich, 156, 157

Russian fleet destruction, 127–28

Saratoga, 40, 47, 56, 91, 102, 106–7, 152; conversion to aircraft carrier, 99–101; and correction to design of flying deck cruisers, 110

Schofield, Frank H., 23, 34, 35, 36–38; on aviation as solution, 90; on floating dock design, 137; on importance of naval aviation, 104; as opposition to flying deck cruisers, 122; questioning of naval air and fleet structure, 102; submarine recommendation by, 49

sea power problem, 29

Seeckt, Hans Von, 3–4

serials (General Board "studies"), 17, 18–21, 41–42, 46, 47

Sims, William S., 9–10, 31, 34, 174–75

South Carolina, 75

Standley, William, 106, 140

Stimson, Henry, 54

Strategic Problem IV, 84–85, 90

Strauss, Joseph, 79–80, 94, 133

submarines: linked to aircrafts, 90, 91; retaining to sink battleships, 68; significance in Orange War Plan, 49; significance to Japan and U.S., 28

submersible fuel tender, 49

Taylor, 115, 116

tenders, 91; budget considerations, 95, 96, 97; as component of building program in Naval Policy, 93; in context of Asiatic Fleet, 92–98, 95–98; and treaty concerns, 93

Tennessee, 64

Tentative Manual for Landing Operations, 131

Terraine, John, 146

Texas, 71–72, 73, 75

"thrusting" strategy of the Orange Plan, 87, 90, 127, 128, 132, 133

Togo (Japanese admiral), 27

Tomosaburo, Kato, 26, 27–28

Tone class (Japan), 105

Torpedo Squadron Twenty (VT-20), 92, 94

Towers, John, 113–14

Treaty of Versailles, 57

treaty system. *See also* fortification clause (Article XIX); General Board of the Navy: "balanced" fleet to support war plans, 52; collaboration in Navy due to the, 9; overview, 1, 8–9; summary of interrelated factors of influence, 168–75; summary of turret modification impact, 82; as a systemic and continuous process, 5–6

Turner, Richmond K., 15, 108; on the Asiatic Fleet naval air situation, 95–97; on the flying deck cruiser, 110–11, 113, 114–15, 119

U-boats, 155–56, 158–59

U.S. Naval Institute, 63
U.S. Naval Policy, 33–35, 36–38, 38, 44, 163. *See also* Naval Air Policy; "blueprints," 34–35, 35–36, 38, 40; "Building and Maintenance Policy" of the, 43–44; General Board serials or "studies," 17, 18–21, 41–42, 46, 47; language relating to battleship modernization, 65; tenders as building component in, 93
U.S. Navy. *See also* battleships; sea power: bureau abbreviations, 11; competing military advice to secretary of, 12; "efficiency" in the, 66; gun range deficiency problem, 75–76, 77–78; larger role in nineteenth and twentieth centuries, 86; officers, culture and mindset of, 9–10, 23–24, 28–29, 67, 86; speed deficiency problem, 75–76, 77; summary of technological solutions, 175; treaty fleet program plan, 65–66
The U.S. Naval Institute *Proceedings,* 30, 130
Utah, 73, 79

Van Keuren, Alexander H. (later Rear Admiral), 68, 109–10; on flying deck cruiser design, 109–10, 113
Versailles, Treaty of, 145–48, 152–56, 159, 173
Virginia, 75

War Plans Division, 13, 14, 31, 38. *See also* Chief of Naval Operations office (CNO or OpNav); Orange War Plan; Schofield, Frank H.; on allocating budget for aircraft carriers, 100; on fleet efficiency, 26; recommendation for tender use, 93; in relation to floating dry docks, 126, 130, 132–33, 139–41; study on use of light cruisers, 106
Washington, 74–75
Washington, Thomas, 92
Washington Naval Conference: and capital ship tonnage ratios and prohibitions, 25–26, 28, 32, 33; and limitation ratios disputed by French, 28; "naval holiday" proposed at, 20–21, 26, 32; ratio system for U.S. relative to Japan, 25, 26–27, 28; Rodgers on Treaty changes, 34–35; scrapping of ships proposed at, 26, 32; ship classes permitted under, 9
Washington Naval Treaty, 32; aircraft carrier/ship conversion allowance under, 42–43; budgetary considerations, 35, 42; in cover letter regarding floating dock design, 138; effect on aircraft carriers introduction, 98–101; events leading to, 145–46; and the failure to impose auxiliary tonnage limits, 28–29, 42; implementation by General Board, 33–39; reconstruction clause of the, 65, 70–72, 74, 99, 163–64; ship building permitted under, 9; and tender concerns, 93; turret redesign debates, 78–81; U.S. Navy Policy drafted to conform to, 20
West Virginia, 64, 82
Whiting, Kenneth, 78, 94–95, 113
Wilbur, Curtis, 20–21, 45–46, 82, 101
Williams, Clarence S., 132, 133
Williams, E. M., 56–57, 85
Wilson, George, 99–100
Wilson, Woodrow, 67
Wright, 93
Wyoming, 73

Yard Floating Docks (YFDs), 130, 136, 141, 142
yardstick, 53
Yarnell, Harry E., 15, 114, 115; in connection with Office of Naval Intelligence studies, 130–31, 134, 140, 166
"yellow" draft studies, 17, 19
YFD-2 ("New Orleans dock"), 130, 141, 142
Yorktown class carriers, 122

Z plan, 154, 155, 156–58

About the Author

Commander John T. Kuehn is a retired naval aviator who flew in both the EP-3 and ES-3 aircraft. He has a wealth of operational experience with almost 200 arrested carrier landings and 3000 flight hours in every operational theater that the United States Navy deployed to. He completed cruises aboard the aircraft carriers USS *Coral Sea,* USS *America,* USS *John F. Kennedy*, and USS *John C. Stennis* and is a combat veteran of both the first Gulf War and the 1995 NATO air campaign over Bosnia. He has served on the faculty of the U.S. Army Command and General Staff College since July 2000 and received his Ph.D. in History from Kansas State University in May 2007. He was recently promoted to Associate Professor of Military History in August 2007. Commander Kuehn has published numerous articles and book reviews and has another book that will also be published this fall, *Eyewitness Pacific Theater* (with Dennis Giangreco).

The Naval Institute Press is the book-publishing arm of the U.S. Naval Institute, a private, nonprofit, membership society for sea service professionals and others who share an interest in naval and maritime affairs. Established in 1873 at the U.S. Naval Academy in Annapolis, Maryland, where its offices remain today, the Naval Institute has members worldwide.

Members of the Naval Institute support the education programs of the society and receive the influential monthly magazine *Proceedings* or the colorful bimonthly magazine *Naval History* and discounts on fine nautical prints and on ship and aircraft photos. They also have access to the transcripts of the Institute's Oral History Program and get discounted admission to any of the Institute-sponsored seminars offered around the country.

The Naval Institute's book-publishing program, begun in 1898 with basic guides to naval practices, has broadened its scope to include books of more general interest. Now the Naval Institute Press publishes about seventy titles each year, ranging from how-to books on boating and navigation to battle histories, biographies, ship and aircraft guides, and novels. Institute members receive significant discounts on the Press's more than eight hundred books in print.

Full-time students are eligible for special half-price membership rates. Life memberships are also available.

For a free catalog describing Naval Institute Press books currently available, and for further information about joining the U.S. Naval Institute, please write to:

Member Services
U.S. Naval Institute
291 Wood Road
Annapolis, MD 21402-5034
Telephone: (800) 233-8764
Fax: (410) 571-1703
Web address: www.usni.org